Disorders of the Human Adrenal Cortex

Endocrine Development

Vol. 13

Series Editor

P. Mullis Bern

Disorders of the Human Adrenal Cortex

Volume Editors

Christa E. Flück Bern
Walter L. Miller San Francisco, Calif.

39 figures, 7 in color, and 8 tables, 2008

KARGER

Basel · Freiburg · Paris · London · New York · Bangalore · Bangkok · Shanghai · Singapore · Tokyo · Sydney

Christa E. Flück
Pediatric Endocrinology and Diabetology
University Children's Hospital
Inselspital
Bern, Switzerland

Walter L. Miller
Department of Pediatrics
University of California
San Francisco, Calif., USA

Library of Congress Cataloging-in-Publication Data

Disorders of the human adrenal cortex / volume editors, C.E. Flück, Walter L. Miller.
p. ; cm. – (Endocrine development, ISSN 1421–7082 ; v. 13)
Includes bibliographical references and index.
ISBN 978-3-8055-8580-4 (hard cover : alk. paper)
1. Adrenal cortex–Diseases. I. Flück, C.E. (Christa E.) II. Miller, Walter L. III. Series.
[DNLM: 1. Adrenal Cortex Diseases. W1 EN3635 v.13 2008/WK 760 D612 2008]
RC659.D57 2008
616.4'5–dc22
2008017636

Bibliographic Indices. This publication is listed in bibliographic services, including Current Contents®

www.karger.com
Printed in Switzerland on acid-free and non-aging paper (ISO 9706) by Reinhardt Druck, Basel
ISSN 1421–7082
ISBN 978–3–8055–8580–4

Contents

Preface

The human adrenals are large and relatively unimportant during fetal life but small and important postnatally. During development, the adrenal cortex undergoes morphological and functional changes that are still not fully understood and may lead to adrenal disease when disordered.

In this volume, the series *Endocrine Development* covers diseases of the human adrenal cortex for the second time. In 2000, volume 2 of this series, edited by Ieuan A. Hughes and Adrian J.L. Clark, brought together a group of experts reviewing clinical and molecular aspects in the volume *Adrenal Disease in Childhood*. Substantial further advances in our understanding of adrenal development, steroid biosynthesis and adrenocortical disorders now necessitate another look at this small but complex organ. Studies of families with adrenal hypoplasia congenita have broadened our knowledge on specific factors involved in the adrenal development. Novel insights into the zonation of the adrenal cortex and adrenal androgen production throughout life have been gained from studies of nonhuman primates. Detailed studies of steroidogenesis in the tammar wallaby pouch young revealed an alternate ('backdoor') pathway to dihydrotestosterone production that is relevant to P450 oxidoreductase deficiency, polycystic ovarian disease, and congenital adrenal hyperplasia (CAH). Finding that mutations in the gene for P450 oxidoreductase cause a complex defect of 17α-hydroxylase and 21-hydroxylase deficiency has defined a new form of CAH and highlighted the pivotal role of electron transfer partners in the activities of steroidogenic enzymes. Critical review of the long-term outcome of prenatal dexamethasone treatment of fetuses at risk for CAH has revealed a potential risk for adverse effects on metabolism, cognitive functions and behavior in later life. Genetic studies of ACTH resistance syndromes and adrenal Cushing's syndrome have determined the causes in some more patients, but have also shown us that there are many more unsolved cases that apparently represent disorders in unknown genes. Finally, showing that adrenal steroidogenesis is widely important, two experts summarize novel aspects of adrenal steroid production in arterial hypertension and the polycystic ovary syndrome.

This book combines ten review chapters written by basic, translational and clinical scientists. Although we tried to cover the newest information gained in the past 5–10 years, there certainly are other developing areas of research concerning the human adrenal cortex. The series *Endocrine Development* does not intend to replace standard endocrine textbooks, and allows the editors to pick a limited number of topics and permits the authors to express their personal opinions.

We thank Primus E. Mullis for inviting us to design this new book on the development and disorders of the human adrenal cortex. Also, we would like to thank all the co-authors for their enthusiasm and effort in sharing their invaluable expertise. Finally, we thank Karger Publishers for bringing this book to the community.

Christa E. Flück
Walter L. Miller

Flück CE, Miller WL (eds): Disorders of the Human Adrenal Cortex.
Endocr Dev. Basel, Karger, 2008, vol 13, pp 1–18

Steroidogenic Enzymes

Walter L. Miller

Division of Endocrinology, Department of Pediatrics, University of California, San Francisco, Calif., USA

Abstract

The enzymes and pathways of steroidogenesis are familiar to most endocrinologists, but the biochemistry and molecular biology of these processes are still being studied. This chapter outlines current knowledge about each enzyme. The quantitative regulation of steroidogenesis occurs at the first step, the conversion of cholesterol to pregnenolone. Chronic regulation is principally at the level of transcription of the gene for P450 side chain cleave (P450scc), which is the enzymatically rate-limiting step. Acute regulation is mediated by steroidogenic acute regulatory protein, which facilitates the rapid influx of cholesterol into mitochondria, where P450scc resides. Qualitative regulation, determining the class of steroid produced, is principally determined by P450c17. In the absence of P450c17 in the zona glomerulosa, C21 deoxy steroids are produced, leading to the mineralocorticoid aldosterone. In the presence of the 17α-hydroxylase but not the 17,20 lyase activity of P450c17 in the zona fasciculata, C21, 17-hydroxy steroids are produced, leading to the glucocorticoid cortisol. When both the 17α-hydroxylase and 17,20 lyase activities of P450c17 are present in the zona reticularis, the androgen precursor dehydroepiandrosterone is produced. The discrimination between 17α-hydroxylase and 17,20 lyase activities is regulated by two posttranslational events, the serine phosphorylation of P450c17 and the allosteric action of cytochrome b5, both of which act to optimize the interaction of P450c17 with its obligatory electron donor, P450 oxidoreductase.

Cholesterol Uptake, Storage, and Transport

The substrate for steroidogenesis is cholesterol (fig. 1). The human adrenal can synthesize cholesterol de novo from acetate, but most of its supply of cholesterol comes from plasma low-density lipoproteins (LDL) derived from dietary cholesterol. Rodent adrenals derive most of their cholesterol from high-density lipoproteins via a receptor termed SR-B1, but this pathway appears to play a minor role in human steroidogenesis [1]. Adequate concentrations of LDL will suppress 3-hydroxy-3-methylglutaryl coenzyme A (HMGCoA) reductase, the rate-limiting enzyme in cholesterol synthesis. Adrenocorticotrophic hormone (ACTH), which stimulates adrenal steroidogenesis, also stimulates the activity of HMGCoA reductase, LDL receptors, and uptake of LDL cholesterol. LDL cholesterol esters are taken up by receptor-mediated

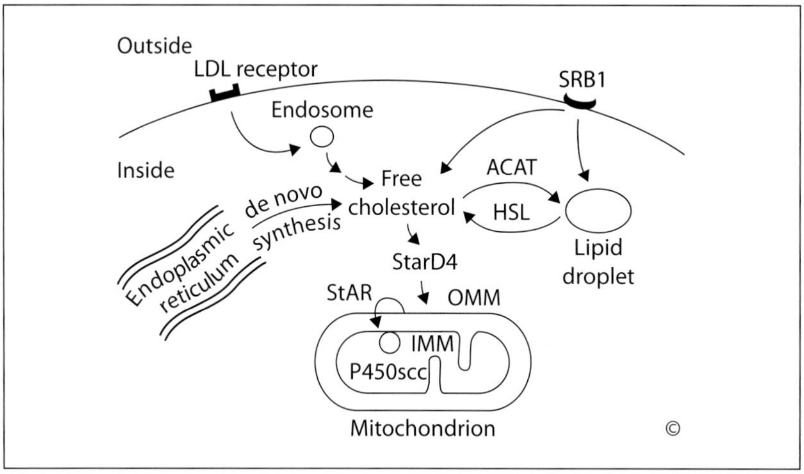

Fig. 1. Cellular cholesterol flux. LDL is picked up by cell-surface receptors in clathrin-coated pits, whereas HDL binds to SR-B1; cholesterol can also be synthesized de novo from acetate in the endoplasmic reticulum. Cholesterol is esterified by ACAT and stored in lipid droplets as cholesterol esters. HSL liberates free cholesterol, which is probably bound by StarD4 and StarD5 proteins for transport to membrane destinations, including the outer mitochondrial membrane (OMM). In the adrenals and gonads, StAR speeds the movement of cholesterol from the OMM to the inner mitochondrial membrane (IMM), where it is converted to pregnenolone by the cholesterol side chain cleavage enzyme, P450scc.

endocytosis, then are stored directly or converted to free cholesterol and used for steroid hormone synthesis. Cholesterol can be esterified by acyl-CoA:cholesterol transferase (ACAT), stored in lipid droplets, and accessed by activation of hormone-sensitive lipase (HSL). ACTH stimulates HSL and inhibits ACAT, thus increasing the availability of free cholesterol for steroid hormone synthesis.

Cytochrome P450 Enzymes

Most steroidogenic enzymes are members of the cytochrome P450 group of oxidases. 'Cytochrome P450' is a generic term for a group of oxidative enzymes, all of which have about 500 amino acids and contain a single heme group. They are termed P450 (pigment 450) because all absorb light at 450 nm in their reduced states. The human genome contains genes for 57 cytochrome P450 enzymes, of which 7 are targeted to mitochondria and 50 are targeted to the endoplasmic reticulum, especially in the liver, where they metabolize toxins, drugs, xenobiotics, and environmental pollutants. Each P450 enzyme can metabolize multiple substrates, catalyzing a broad array of oxidations.

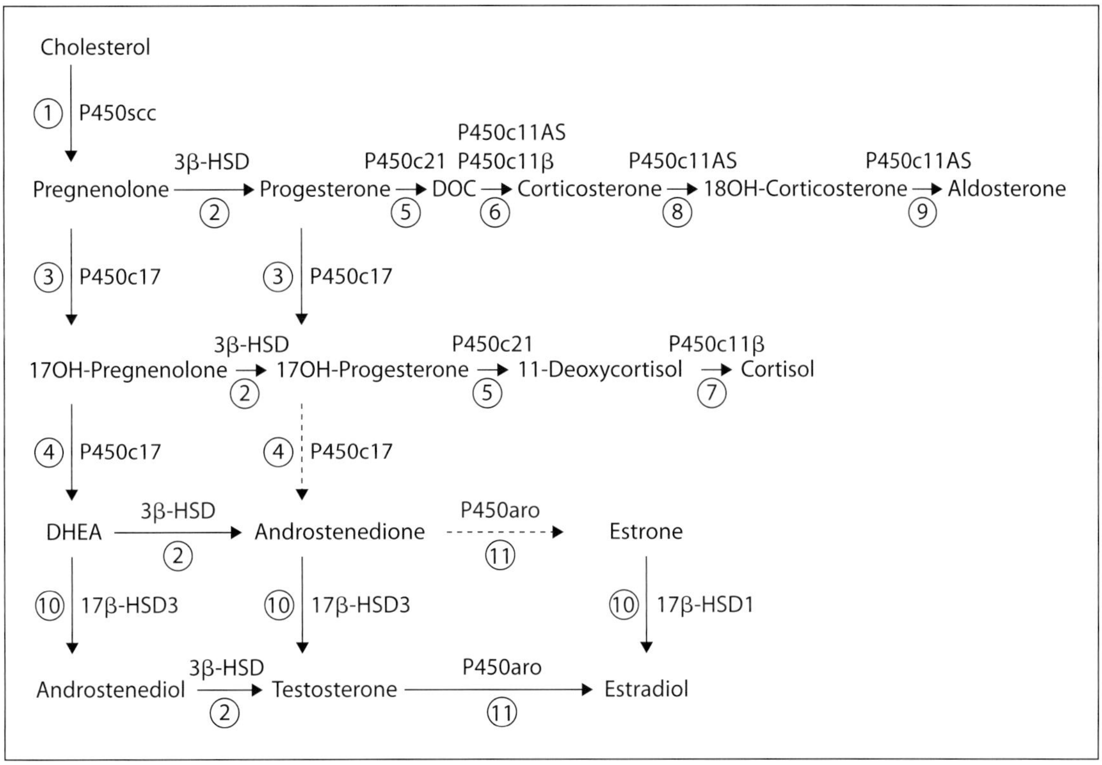

Fig. 2. Principal pathways of human steroidogenesis. The names of the enzymes are shown by each reaction, and the traditional names of the enzymatic activities correspond to the circled numbers. (1) Mitochondrial cytochrome P450scc catalyzes 20α-hydroxylation, 22-hydroxylation, and scission of the C20–22 carbon bond. (2) 3β-HSD catalyzes 3β-hydroxysteroid dehydrogenase and isomerase activities, converting Δ^5 steroids to Δ^4 steroids. (3) P450c17 catalyzes the 17α-hydroxylation of pregnenolone to 17OH-pregnenolone and of progesterone to 17OH-progesterone. (4) P450c17 catalyzes 17,20 lyase activity, converting 17OH-pregnenolone to DHEA; only insignificant amounts of 17OH-progesterone are converted to Δ^4 androstenedione by human P450c17, although this reaction occurs in other species. (5) P450c21 catalyzes the 21-hydroxylation of progesterone to DOC and of 17OH-progesterone to 11-deoxycortisol. (6) DOC is converted to corticosterone by the 11-hydroxylase activity of P450c11AS in the zona glomerulosa and by P450c11β in the zona fasciculata. (7) 11-deoxycortisol undergoes 11β-hydroxylation by P450c11β to produce cortisol in the zona fasciculata. (8, 9) P450c11AS catalyzes 18-hydroxylase and 18-methyl oxidase activities, converting corticosterone to 18OH-corticosterone and aldosterone in the zona glomerulosa. (10) In testis, 17β-HSD3 converts DHEA to androstenediol and androstenedione to testosterone; in placenta and ovary, 17β-HSD1 converts estrone to estradiol. (11) P450aro aromatizes testosterone to estradiol and aromatizes androstenedione to estrone.

Five distinct P450 enzymes are involved in adrenal steroidogenesis (fig. 2). Mitochondrial P450scc is the cholesterol side-chain cleavage enzyme catalyzing the series of reactions formerly termed '20,22 desmolase'. Two mitochondrial isozymes of P450c11, P450c11β and P450c11AS, catalyze 11β-hydroxylase, 18-hydroxylase, and

18-methyl oxidase activities. In the endoplasmic reticulum, P450c17 catalyzes both 17α-hydroxylase and 17,20 lyase activities, and P450c21 catalyzes the 21-hydroxylation of both glucocorticoids and mineralocorticoids. In the gonads and elsewhere, P450aro in the endoplasmic reticulum catalyzes aromatization of androgens to estrogens.

Hydroxysteroid Dehydrogenases

The hydroxysteroid dehydrogenases have molecular masses of about 35–45 kDa, do not have heme groups, and require NAD^+ or $NADP^+$ as cofactors. Whereas most steroidogenic reactions catalyzed by P450 enzymes are due to the action of a single form of P450, each of the reactions catalyzed by hydroxysteroid dehydrogenases can be catalyzed by at least two, often very different, isozymes. Members of this family include the 3α- and 3β-hydroxysteroid dehydrogenases, the two 11β-hydroxysteroid dehydrogenases, and a series of 17β-hydroxysteroid dehydrogenases; the 5α-reductases are unrelated to this family. Based on their structures, these enzymes fall into two groups: the short-chain dehydrogenase reductase (SDR) family, characterized by a 'Rossman fold', and the aldo-keto reductase (AKR) family, characterized by a triosephosphate isomerase barrel motif [2]. The SDR enzymes include 11β-HSDs 1 and 2, and 17β-HSDs 1, 2, 3, and 4; the AKR enzymes include 17β-HSD5, which is important in extraglandular activation of androgenic precursors. It is physiologically more useful to classify these enzymes as dehydrogenases or reductases. The dehydrogenases use NAD^+ as their cofactor to oxidize hydroxysteroids to ketosteroids, and the reductases mainly use NADPH to reduce ketosteroids to hydroxysteroids. Although these enzymes are typically bidirectional in vitro, they tend to function in only one direction in intact cells, with the direction determined by the cofactor(s) available [2].

P450scc and the Chronic Regulation of Steroidogenesis

Conversion of cholesterol to pregnenolone in mitochondria is the first, rate-limiting and hormonally regulated step in the synthesis of all steroid hormones. This involves three distinct chemical reactions, 20α-hydroxylation, 22-hydroxylation, and scission of the cholesterol side chain to yield pregnenolone and isocaproic acid. These three sequential reactions are catalyzed by P450scc, encoded by a single gene on chromosome 15 [3]. A cell is said to be 'steroidogenic' if it expresses P450scc, and thus is able to convert cholesterol to pregnenolone; some 'peripheral' cell types that lack P450scc (e.g. hepatocytes and adipocytes) can modify circulating steroids, but are unable to synthesize steroids de novo. Deletion of the gene for P450scc in the rabbit [4] or mouse [5] eliminates all steroidogenesis, demonstrating that P450scc is the only enzyme that can produce pregnenolone in vivo.

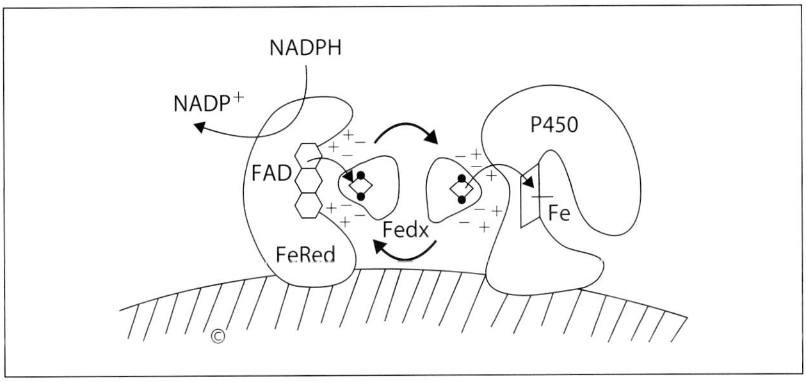

Fig. 3. Electron transport to a mitochondrial cytochrome P450. Ferredoxin reductase (FeRed) accepts electrons from NADPH, converting it to $NADP^+$. The electrons are passed to ferredoxin (Fedx), which functions as a freely diffusable electron shuttle mechanism. Electrons from charged ferredoxin are accepted by any available cytochrome P450. The uncharged ferredoxin may then be again bound to ferredoxin reductase to receive another pair of electrons. For P450scc, three pairs of electrons must be transported to the P450 to convert cholesterol to pregnenolone.

Two quantitative regulatory mechanisms – acute and chronic – determine the amount of steroid a steroidogenic cell will produce. The transcription of genes encoding steroidogenic enzymes, principally P450scc, determines the steroidogenic capacity of a cell by determining the amount of each steroidogenic enzyme present. P450scc is a very slow enzyme, converting only ~6 molecules of cholesterol to pregnenolone per molecule of P450scc per minute [6]. As P450scc is the rate-limiting enzyme in steroid production, the regulation of its gene determines the amount of steroid a cell can produce – this is the chronic regulation of steroidogenesis. By contrast, acute regulation determines the amount of steroid produced in response to provocative stimuli. This action is mediated by the steroidogenic acute regulatory protein (StAR), which facilitates the movement of cholesterol from the outer to the inner mitochondrial membrane, where it can be acted on by P450scc [7].

Transport of Electrons to P450scc: Ferredoxin Reductase and Ferredoxin

P450scc functions as the terminal oxidase in a mitochondrial electron transport system [8]. Electrons from NADPH are accepted by a flavoprotein, termed ferredoxin reductase, that is loosely associated with the inner mitochondrial membrane. Ferredoxin reductase transfers the electrons to an iron/sulfur protein termed ferredoxin, which is found in the mitochondrial matrix or loosely adherent to the inner mitochondrial membrane. Ferredoxin then transfers the electrons to P450scc (fig. 3).

Ferredoxin reductase and ferredoxin serve as electron transfer proteins for all mitochondrial P450s. Ferredoxin forms a 1:1 complex with ferredoxin reductase, then dissociates, then subsequently reforms an analogous 1:1 complex with a mitochondrial P450 such as P450scc or P450c11, thus functioning as an indiscriminate electron shuttle mechanism. Ferredoxin reductase is a membrane-bound mitochondrial flavoprotein that receives electrons from NADPH. The human genes for ferredoxin reductase and ferredoxin are expressed in all tissues, suggesting they may also have other roles [8]. Human mutations in these genes have not been described.

The Steroidogenic Acute Regulatory Protein and the Acute Regulation of Steroidogenesis

Acute regulation, where steroids are released within minutes of a stimulus, is at the level of cholesterol access to P450scc [7, 9]. When steroidogenic cells are treated with inhibitors of protein synthesis such as cycloheximide, the acute steroidogenic response is eliminated, indicating that a short-lived, cycloheximide-sensitive protein triggers the acute steroidogenic response. This factor, first identified as 30- and 37-kDa phosphoproteins that were rapidly synthesized when steroidogenic cells were stimulated with tropic hormones, was cloned from mouse Leydig MA-10 cells and named the steroidogenic acute regulatory protein [10]. The central role of StAR in steroidogenesis was proven by finding that mutations of StAR caused congenital lipoid adrenal hyperplasia [11, 12].

Some adrenal steroidogenesis is independent of StAR; when nonsteroidogenic cells are transfected with StAR and the P450scc system, they convert cholesterol to pregnenolone at about 14% of the StAR-induced rate [11, 12]. Furthermore, the placenta utilizes mitochondrial P450scc to initiate steroidogenesis but does not express StAR [13]. The mechanism of StAR-independent steroidogenesis is unclear; it may occur without a triggering protein, or some other protein may exert StAR-like activity to promote cholesterol flux, but without StAR's rapid kinetics. Substantial data indicate that the action of StAR also requires the peripheral benzodiazepine receptor on the outer mitochondrial membrane [14–16]. Recent data utilizing cross-linking of StAR to isolated mitochondria and mass spectrometric analysis of the cross-linked complexes show that StAR's action requires interaction with two proteins on the outer mitochondrial membrane – voltage-dependent anion channel 1 and phosphate carrier protein [17]. The details of the mechanism of StAR's action remain unclear, but two facts are now well established. First, StAR acts exclusively on the outer mitochondrial membrane and does not need to enter the mitochondria to be active [7, 18]. Second, the interaction with the outer mitochondrial membrane induces structural changes, characterized as a molten globule transition, that are required for StAR's activity, probably permitting it to take up and discharge cholesterol [7, 19, 20].

3β-Hydroxysteroid Dehydrogenase/$\Delta^5 \rightarrow \Delta^4$ Isomerase

Once pregnenolone is produced from cholesterol, it may undergo 17α-hydroxylation by P450c17 to yield 17-hydroxypregnenolone, or it may be converted to progesterone, the first biologically important steroid in the pathway. A single 42-kDa microsomal enzyme, 3β-hydroxysteroid dehydrogenase (3β-HSD) catalyzes both conversion of the hydroxyl group to a keto group on carbon 3 and the isomerization of the double bond from the B ring (Δ^5 steroids) to the A ring (Δ^4 steroids) [21–23]. 3β-HSD converts pregnenolone to progesterone, 17α-hydroxypregnenolone to 17α-hydroxyprogesterone (17OHP), dehydroepiandrosterone (DHEA) to androstenedione, and androstenediol to testosterone, with very similar K_m and V_{max} values [24]. There are two isozymes of 3β-HSD, encoded by closely linked, evolutionarily duplicated genes, that have 93.5% amino acid sequence identity and are enzymatically very similar. 3β-HSD2 is expressed in the adrenals and gonads, while 3β-HSD1 is expressed in placenta, breast, and 'extraglandular' tissues. Bovine 3β-HSD is found in both the endoplasmic reticulum and in mitochondria [25], suggesting that subcellular distribution may provide a novel point regulating the direction of steroidogenesis.

P450c17 and the Qualitative Regulation of Steroidogenesis

Both pregnenolone and progesterone may undergo 17α-hydroxylation to 17α-hydroxypregnenolone and 17OHP, respectively. 17OHP may also undergo scission of the C17,20 carbon bond to yield DHEA; however, very little 17OHP is converted to androstenedione because the human P450c17 enzyme catalyzes this reaction at only 3% of the rate for conversion of 17α-hydroxypregnenolone to DHEA [26]. These reactions are all mediated by a single enzyme, P450c17. This P450 is bound to the endoplasmic reticulum, where it accepts electrons from P450 oxidoreductase (POR). As P450c17 has both 17α-hydroxylase activity and 17,20 lyase activity, it is the key branch point in steroid hormone synthesis. In the absence of P450c17, a steroidogenic cell produces C21 17-deoxysteroids (e.g. progesterone in the ovarian granulosa cell or aldosterone in the adrenal glomerulosa cell). If only the 17α-hydroxylase activity of P450c17 is present (e.g. in the adrenal zona fasiculata), C21 17-hydroxysteroids (e.g. cortisol) are produced. If both the 17α-hydroxylase and 17,20-lyase activities of P450c17 are present (e.g. in ovarian theca cells, testicular Leydig cells, or adrenal zona reticularis), C19 precursors of sex steroids (e.g. DHEA) are produced (fig. 2).

17α-hydroxylase and 17,20 lyase were once thought to be separate enzymes. The adrenals of prepubertal children synthesize ample cortisol but virtually no sex steroids (i.e. have 17α-hydroxylase activity but not 17,20 lyase activity), until adrenarche initiates production of adrenal androgens (i.e. turns on 17,20 lyase activity).

Furthermore, patients had been described lacking 17,20 lyase activity but retaining normal 17α-hydroxylase activity. However, studies of pig P450c17 showed that both 17α-hydroxylase and 17,20 lyase activities are catalyzed by a single protein [27], and cells transfected with a vector expressing P450c17 cDNA acquire both 17α-hydroxylase and 17,20 lyase activities [28]. P450c17 is encoded by a single gene on chromosome 10q24.3 [29] that is structurally related to the genes for P450c21 (21-hydroxylase) [30].

Thus, the distinction between 17α-hydroxylase and 17,20 lyase is functional and not genetic or structural. Human P450c17 catalyzes 17α-hydroxylation of pregnenolone and progesterone equally well, but the 17,20 lyase activity of human P450c17 strongly prefers 17OH pregnenolone and not 17OH progesterone, consistent with the large amounts of DHEA secreted by both the fetal and adult adrenal. Furthermore, the 17α-hydroxylase reaction occurs more readily than the 17,20 lyase reaction. The ratio of the 17α-hydroxylase to 17,20-lyase activity of P450c17 determines the ratio of cortisol to DHEA produced by the adrenal. This ratio differs in the fasiculata and reticularis, and is developmentally regulated during adrenarche. The regulation of 17,20 lyase activity is mediated by three factors that affect P450c17 post-translationally: (a) the abundance of the electron-donating protein POR [31, 32], (b) the allosteric action of cytochrome b5 [26], and (c) the serine phosphorylation of P450c17 [33–35]. These three factors all act by the same mechanism – increasing the efficiency of electron transfer from POR.

Electron Transport to P450c17: P450 Oxidoreductase and Cytochrome b5

P450c17 (and P450c21) receive electrons from POR, a membrane-bound flavoprotein, distinct from the mitochondrial flavoprotein ferredoxin reductase [8]. POR receives two electrons from NADPH and transfers them one at a time to the P450 [36]. Electron transfer for the lyase reaction is promoted by the action of cytochrome b5 as an allosteric factor rather than as an alternate electron donor [26]. 17,20 lyase activity also requires the phosphorylation of serine residues on P450c17 by a cAMP-dependent protein kinase [33–35] (fig. 4). The availability of electrons determines whether P450c17 performs only 17α-hydroxylation, or also performs 17,20 bond scission; increasing the ratio of POR or cytochrome b5 to P450c17 in vitro or in vivo increases the ratio of 17,20 lyase activity to 17α-hydroxylase activity. Competition between P450c17 and P450c21 for available 17OHP does not appear to be important in determining whether 17OHP undergoes 21-hydroxylation or 17,20 bond scission [32]. Thus, the regulation of 17,20 lyase activity, and consequently of DHEA production, depends on factors that facilitate the flow of electrons to P450c17: high concentrations of POR, the presence of cytochrome b5, and serine phosphorylation of P450c17.

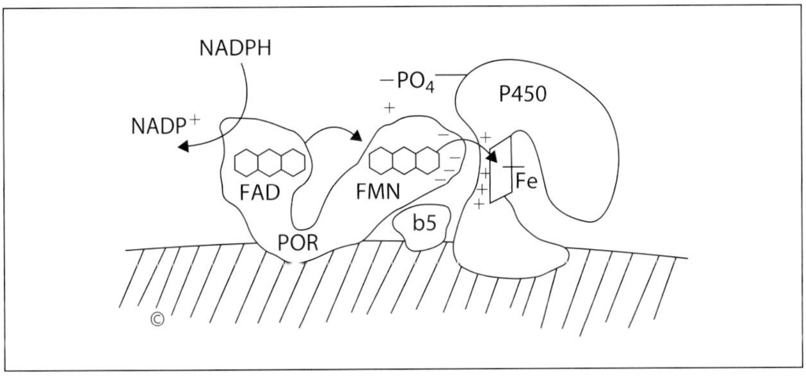

Fig. 4. Electron transport to a microsomal cytochrome P450. NADPH interacts with POR and gives up a pair of electrons, which are received by the flavin adenine dinucleotide (FAD) moiety. Electron receipt elicits a conformational change, permitting the isoalloxazine rings of the FAD and flavin mononucleotide (FMN) moieties to come close together, so that the electrons pass from the FAD to the FMN. Following a second conformational change that returns the protein to its original orientation, the FMN domain of POR interacts with the P450 and electrons from the FMN domain of POR reach the heme group to mediate catalysis. The interaction of POR and the P450 is coordinated by negatively charged acidic residues on the surface of the FMN domain of POR, and positively charged basic residues in the concave redox-partner binding site of the P450. In the case of human P450c17, this interaction is facilitated by the allosteric action of cytochrome b5, and by the serine phosphorylation of P450c17.

P450c21

Progesterone and 17OHP can be 21-hydroxylated at to yield deoxycorticosterone (DOC) and 11-deoxycortisol, respectively (fig. 2). The nature of 21-hydroxylation has been of great clinical interest because disordered 21-hydroxylation causes more than 90% of congenital adrenal hyperplasia (CAH). The clinical symptoms associated with CAH are complex and devastating. Decreased cortisol and aldosterone synthesis can lead to sodium loss, potassium retention, and hypotension, which will lead to cardiovascular collapse and death, usually within a month after birth if not treated appropriately. Decreased synthesis of cortisol in utero leads to overproduction of ACTH and consequent overstimulation of adrenal steroid synthesis; 17OHP accumulates because P450c17 converts 17OHP to androstenedione very inefficiently. However, 17-hydroxypregnenolone also accumulates and is converted to DHEA, and subsequently to androstenedione and testosterone, resulting in severe prenatal virilization of female fetuses. Although some patients appear to have a 21-hydroxylating defect confined to the zona fasiculata, there is only one 21-hydroxylase encoded by a single functional gene on chromosome 6p21 [37–39]. As this gene lies in the middle of the major histocompatibility locus, disorders of adrenal 21-hydroxylation are closely linked to specific HLA types.

Adrenal 21-hydroxylation is mediated by P450c21 found in smooth endoplasmic reticulum. P450c21 employs the same POR used by P450c17 to transport electrons from NADPH. 21-hydroxylase activity has also been described in a broad range of adult and fetal extra-adrenal tissues [40], but this activity is not mediated by the P450c21 enzyme found in the adrenal [41]; at least three hepatic P450 enzymes can catalyze 21-hydroxylation in vitro [42], but the clinical significance of these activities in clinical situations is unclear. Thus, patients with severe P450c21 mutations may still have appreciable plasma concentrations of 21-hydroxylated steroids.

P450c11β and P450c11AS

P450c11β and P450c11AS, two isozymes that have 93% amino acid sequence identity and are encoded by tandemly duplicated genes on chromosome 8q21–22, catalyze the final steps in the synthesis of both glucocorticoids and mineralocorticoids [43]. Both forms of P450c11 are found on the inner mitochondrial membrane, and use ferredoxin and ferredoxin reductase to receive electrons from NADPH. By far the more abundant of the two isozymes is P450c11β, the 11β-hydroxylase that converts 11-deoxycortisol to cortisol and 11-DOC to corticosterone in the zona fasciculata. The less abundant isozyme, P450c11AS, is found only in the zona glomerulosa, where it has 11β-hydroxylase, 18-hydroxylase and 18-methyl oxidase (aldosterone synthase) activities; thus P450c11AS is able to catalyze all the reactions needed to convert DOC to aldosterone [43].

P450c11β is encoded by the CYP11B1 gene, which is induced by ACTH and suppressed by glucocorticoids. P450c11AS is encoded by the CYP11B2 gene, which is induced by angiotensin II and potassium ion. Patients with disorders in P450c11β have classical 11β-hydroxylase deficiency but can still produce aldosterone, while patients with disorders in P450c11AS have rare forms of aldosterone deficiency (so-called corticosterone methyl oxidase deficiency) while retaining the ability to produce cortisol [44–46].

17β-Hydroxysteroid Dehydrogenases

Androstenedione is converted to testosterone, DHEA is converted to androstenediol, and estrone is converted to estradiol by the 17β-hydroxysteroid dehydrogenases (17β-HSD), [47]. The 17β-HSDs can be confusing because: (a) there are several different 17β-HSDs, (b) some are preferential oxidases while others are preferential reductases, (c) they differ in their substrate preference and sites of expression, (d) there is inconsistent nomenclature, especially with the rodent enzymes, and (e) some proteins termed 17β-HSD actually have very little 17β-HSD activity, and are principally involved in other reactions [48].

Type 1 17β-HSD (17β-HSD1) is exclusively estrogenic. 17β-HSD1 is a 34-kDa cytosolic reductive SDR enzyme first isolated and cloned from the placenta, where it produces estriol, and is expressed in ovarian granulosa cells, where it produces estradiol [22, 49, 50]. 17β-HSD1 uses NADPH as its cofactor to catalyze reductase activity.

It acts as a dimer and only accepts steroid substrates with an aromatic A ring, so that its activity is confined to activating estrogens. The 3-dimensional structure of human 17β-HSD1 has been determined by X-ray crystallography [51]. No genetic deficiency syndrome for 17β-HSD1 has been described.

17β-HSD2 is a microsomal oxidase that uses NAD^+ to inactivate both estradiol to estrone and testosterone to Δ^4 androstenedione. 17β-HSD2 is found in the placenta, liver, small intestine, prostate, secretory endometrium and ovary. In contrast to 17β-HSD1, which is found in placental syncytiotrophoblast cells, 17β-HSD2 is expressed in endothelial cells of placental intravillous vessels, consistent with its apparent role in defending the fetal circulation from transplacental passage of maternal estradiol or testosterone [52]. No deficiency state for 17β-HSD2 has been reported.

17β-HSD3, the androgenic form of 17β-HSD, is a microsomal enzyme that is apparently expressed only in the testis [53]. This is the enzyme that is disordered in the classic syndrome of male pseudohermaphroditism that is often termed 17-ketosteroid reductase deficiency.

An enzyme termed 17β-HSD4 was initially identified as an NAD^+-dependent oxidase with activities similar to 17β-HSD2 [54], but this peroxisomal protein is primarily an enoyl-CoA hydratase and 3-hydroxyacyl-CoA dehydrogenase [55, 56]. Deficiency of 17β-HSD4 causes a form of Zellweger syndrome, in which bile acid biosynthesis is disturbed but steroidogenesis is not [56].

17β-HSD5, originally cloned as a 3α-hydroxysteroid dehydrogenase [57], is an AKR enzyme that catalyzes the reduction of Δ^4 androstenedione to testosterone [58]. The 17β-HSD activity of 17β-HSD5 is quite labile in vitro [58], hence its activity in androgen biosynthesis has been less clear, but it appears to be responsible for low levels of testosterone synthesis in the adrenal and adipose tissue and may also convert androstenedione to testosterone in muscle.

Steroid Sulfotransferase and Sulfatase

Steroid sulfates may be synthesized directly from cholesterol sulfate or may be formed by sulfation of steroids by cytosolic sulfotransferase (SULT) enzymes [59, 60]. At least 44 distinct isoforms of these enzymes have been identified belonging to five families of SULT genes; many of these genes yield alternately spliced products accounting for the large number of enzymes. The SULT enzymes that sulfonate steroids include SULT1E (estrogens), SULT2A1 (nonaromatic steroids) and SULT2B1 (sterols). SULT2A1 is the principal sulfotransferase expressed in the adrenal, where it sulfates the 3β-hydroxyl group of Δ^5 steroids (pregnenolone, 17OH-pregnenolone, DHEA, androsterone) but not of cholesterol. SULT2B1a will also sulfonate pregnenolone but not cholesterol, whereas cholesterol is the principal substrate for SULT2B1b in the skin, liver and elsewhere. It is not clear whether most steroid sulfates are simply inactivated forms of steroid or if they serve specific hormonal roles. Knockout of the mouse SULT1E1 gene is associated with elevated estrogen levels, increased expression of tissue factor in the placenta, and increased platelet activation, leading to placental thrombi and fetal loss that could be ameliorated by anticoagulant therapy [61]. Mutations ablating the function of human SULT enzymes have not been described, but single nucleotide polymorphisms that alter the amino acid sequences and catalytic activity affecting drug activity are well described. African-Americans have a high rate of polymorphisms in SULT2A1 apparently influencing plasma ratios of DHEA:DHEA sulfate (DHEAS), which may correlate with risk of prostatic and other cancers [62].

Steroid sulfates may also be hydrolyzed to the native steroid by steroid sulfatase. Deletions in the steroid sulfatase gene on chromosome Xp22.3 cause X-linked ichthyosis [63]. In the fetal adrenal and placenta, diminished or absent sulfatase deficiency reduces the pool of free DHEA available for placental conversion to estrogen, resulting in low concentrations of estriol in the maternal blood and urine. The accumulation of steroid sulfates in the stratum corneum of the skin causes the ichthyosis. Steroid sulfatase is also expressed in the fetal rodent brain, possibly converting peripheral DHEAS to active DHEA [64].

Aromatase: P450aro

Estrogens are produced by the aromatization of androgens by a complex series of reactions catalyzed by P450aro [65, 66]. This typical cytochrome P450 is encoded by a single gene on chromosome 15q21.1. This gene uses several different promoter sequences, transcriptional start sites, and alternatively chosen first exons to encode P450aro mRNA in different tissues under different hormonal regulation. P450aro expression in extraglandular tissues, especially fat, can convert adrenal androgens to estrogens. P450aro in the epiphyses of growing bone converts testosterone to estradiol; the tall stature, delayed epiphyseal maturation and osteopenia of males with aromatase deficiency, and their rapid reversal with estrogen replacement indicate that estrogen, not androgen, is responsible for epiphyseal maturation in males [66]. Although it has traditionally been thought that aromatase activity is needed for embryonic and fetal development, infants and adults with genetic disorders in this enzyme have been described, showing that fetoplacental estrogen is not needed for normal fetal development [67].

5α-Reductase

Testosterone is converted to the more potent androgen, dihydrotestosterone, by 5α-reductase in testosterone's target tissues. There are two isozymes of 5α-reductase: type 1, found in the scalp and other peripheral tissues, is encoded by a gene on chromosome 5; type 2, the predominant form found in male reproductive tissues, in encoded by a structurally related gene on chromosome 2p23 [68]. The syndrome of

5α-reductase deficiency, a disorder of male sexual differentiation, is due to mutations in the gene encoding the type 2 enzyme [69]. The type 1 and 2 genes show an unusual pattern of developmental regulation of expression. The type 1 gene is not expressed in the fetus, then is expressed briefly in the skin of the newborn, and then remains unexpressed until its activity and protein are again found after puberty. The type 2 gene is expressed in fetal genital skin, in the normal prostate, and in prostatic hyperplasia and adenocarcinoma. Thus, the type 1 enzyme may be responsible for the pubertal virilization seen in patients with classic 5α-reductase deficiency, and the type 2 enzyme may be involved in male pattern baldness [68].

11β-Hydroxysteroid Dehydrogenases

The interconversion of cortisol and cortisone is mediated by two isozymes of 11β-hydroxysteroid dehydrogenase (11β-HSD), each of which has both oxidase and reductase activity, depending on the cofactor available ($NADP^+$ or NADPH) [70]. The type 1 enzyme (11β-HSD1) is expressed mainly in glucocorticoid-responsive tissues such as the liver, testis, lung and proximal convoluted tubule. 11β-HSD1 can catalyze both the oxidation of cortisol to cortisone using $NADP^+$ as its cofactor (K_m 1.6 μM), or the reduction of cortisone to cortisol using NADPH as its cofactor (K_m 0.14 μM); the reaction catalyzed depends on which cofactor is available, but the enzyme can only function with high (micromolar) concentrations of steroid [71, 72]. 11β-HSD2 catalyzes only the oxidation of cortisol to cortisone using NADH, and can function with low (nanomolar) concentrations of steroid (K_m 10–100 nM) [73]. 11β-HSD2 is expressed in mineralocorticoid-responsive tissues and thus serves to 'defend' the mineralocorticoid receptor by inactivating cortisol to cortisone, so that only 'true' mineralocorticoids, such as aldosterone or DOC can exert a mineralocorticoid effect. Thus 11β-HSD2 prevents cortisol from overwhelming renal mineralocorticoid receptors, and in the placenta and other fetal tissues 11β-HSD2 also inactivates cortisol [74]. The placenta also has abundant $NADP^+$ favoring the oxidative action of 11β-HSD1, so that in placenta both enzymes protect the fetus from high maternal concentrations of cortisol [70]. 11β-HSD1 is located on the luminal side of the endoplasmic reticulum, and hence is not in contact with the cytoplasm. In this unusual cellular location, 11β-HSD1 receives NADPH provided by the action of hexose-6-phosphate dehydrogenase [75]. This links 11β-HSD1 to the pentose monophosphate shunt, providing a direct paracrine link between local glucocorticoid production and energy storage as fat.

Fetal Adrenal Steroidogenesis

Adrenocortical steroidogenesis begins around week 7 of gestation. Steroidogenic enzymes are immunocytochemically detected principally in the fetal zone at 50–52 days after conception, and by 8 weeks after conception the adrenal contains cortisol and responds to ACTH in primary culture systems [76]. This cortisol synthesis is under the regulation of pituitary ACTH and involves transient expression of adrenal 3β-HSD2; following the 9th week after conception, expression of 3β-HSD2 and synthesis of cortisol wane; 3β-HSD2 is barely detectable at 10–11 weeks and is absent at 14 weeks. At the same time, the fetal adrenal also produces 17β-HSD5 [76], which converts androstenedione to testosterone. Thus the fetal adrenal makes cortisol at the same time during gestation that fetal testicular testosterone is virilizing the genitalia of the normal male fetus. This fetal adrenal cortisol apparently suppresses ACTH, which otherwise would drive adrenal testosterone synthesis via 17β-HSD5.

Fetuses affected with genetic lesions in adrenal steroidogenesis can produce sufficient adrenal androgen to virilize a female fetus to a nearly male appearance, and this masculinization of the genitalia is complete by the 12th week of gestation. The definitive zone of the fetal adrenal produces steroid hormones according to the pathways in figure 2. By contrast, the large fetal zone of the adrenal is relatively deficient in 3β-HSD2 activity after 12 weeks. The fetal adrenal has relatively abundant 17,20 lyase activity of P450c17; low 3β-HSD and high 17,20 lyase activity account for the abundant production of DHEA and DHEAS by the fetal adrenal, which are converted to estrogens by the placenta. The fetal adrenal also has considerable sulfotransferase activity but little steroid sulfatase activity, also favoring conversion of DHEA to DHEAS. The resulting DHEAS cannot be a substrate for adrenal 3β-HSD2; instead, it is secreted, 16α-hydroxylated in the fetal liver, and then acted on by placental steroid sulfatase, 3β-HSD1, 17β-HSD1, and P450aro to produce estriol, or the substrates can bypass the liver to yield estrone and estradiol. Placental estrogens inhibit adrenal 3β-HSD activity, providing a feedback system to promote production of DHEAS [77]. Fetal adrenal steroids account for 50% of the estrone and estradiol and 90% of the estriol in the maternal circulation.

Although the fetoplacental unit produces huge amounts of DHEA, DHEAS and estriol, as well as other steroids, they do not appear to serve an essential role. Successful pregnancy depends on placental synthesis of progesterone, which suppresses uterine contractility and prevents spontaneous abortion; however, fetuses with genetic disorders of adrenal and gonadal steroidogenesis develop normally, reach term gestation and undergo normal parturition and delivery. Mineralocorticoid production is only required postnatally, estrogens are not required, androgens are only needed for male sexual differentiation. Human fetal glucocorticoids may be needed at about 8–12 weeks [76], but it is not clear that they are needed thereafter; if they are, the small amount of maternal cortisol that escapes placental inactivation suffices.

The regulation of steroidogenesis and growth of the fetal adrenal are not fully understood, but both are related to ACTH. ACTH effectively stimulates steroidogenesis by fetal adrenal cells in vitro [78], and excess ACTH is clearly involved in the adrenal growth and overproduction of androgens in fetuses affected with CAH. Prenatal treatment of such fetuses by administering pharmacologic doses of dexamethasone to the mother at 6–10 weeks gestation can significantly reduce fetal adrenal androgen production and thus reduce the virilization of female fetuses, thus, the hypothalamic-pituitary-adrenal axis functions very early in fetal life. By contrast, however, anencephalic fetuses lacking pituitary ACTH have adrenals that contain a fairly normal complement of steroidogenic enzymes and retain their capacity for steroidogenesis. Thus, fetal adrenal steroidogenesis may be regulated by both ACTH-dependent and ACTH-independent mechanisms.

Acknowledgement

Supported by National Institutes of Health Grant HD41958.

References

1 Ikonen E: Mechanisms for cellular cholesterol transport: Defects and human disease. Physiol Rev 2006;86:1237–1261.

2 Agarwal AK, Auchus RJ: Cellular redox state regulates hydroxysteroid dehydrogenase activity and intracellular hormone potency. Endocrinology 2005;146:2531–2538.

3 Chung B, Matteson KJ, Voutilainen R, Mohandas TK, Miller WL: Human cholesterol side-chain cleavage enzyme, P450scc: cDNA cloning, assignment of the gene to chromosome 15, and expression in the placenta. Proc Natl Acad Sci USA 1986;83:8962–8966.

4 Yang X, Iwamoto K, Wang M, Artwohl J, Mason JI, Pang S: Inherited congenital adrenal hyperplasia in the rabbit is caused by a deletion in the gene encoding cytochrome P450 cholesterol side-chain cleavage enzyme. Endocrinology 1993;132:1977–1982.

5 Hu MC, Hsu NC, El Hadj NB, Pai CI, Chu HP, Wang CKL, Chung BC: Steroid deficiency syndromes in mice with targeted disruption of Cyp11a1. Mol Endocrinol 2002;16:1943–1950.

6 Kuwada M, Kitajima R, Suzuki H, Horie S: Purification and properties of cytochrome P-450(SCC) from pig testis mitochondria. Biochem Biophy Res Com 1991;176:1501–1508.

7 Miller WL: StAR search–what we know about how the steroidogenic acute regulatory protein mediates mitochondrial cholesterol import. Mol Endocrinol 2007;21:589–601.

8 Miller WL: Regulation of steroidogenesis by electron transfer. Endocrinology 2005;146:2544–2550.

9 Stocco DM, Clark BJ: Regulation of the acute production of steroids in steroidogenic cells. Endocr Rev 1996;17:221–244.

10 Clark BJ, Wells J, King SR, Stocco DM: The purification, cloning and expression of a novel luteinizing hormone-induced mitochondrial protein in MA-10 mouse Leydig tumor cells. Characterization of the steroidogenic acute regulatory protein (StAR). J Biol Chem 1994;269:28314–28322.

11 Lin D, Sugawara T, Strauss JF III, Clark BJ, Stocco DM, Saenger P, Rogol A, Miller WL: Role of steroidogenic acute regulatory protein in adrenal and gonadal steroidogenesis. Science 1995;267:1828–1831.

12 Bose HS, Sugawara T, Strauss JF III, Miller WL: The pathophysiology and genetics of congenital lipoid adrenal hyperplasia. N Engl J Med 1996;335:1870–1878.

13 Sugawara T, Holt JA, Driscoll D, Strauss JF III, Lin D, Miller WL, Patterson D, Clancy KP, Hart IM, Clark BJ, Stocco DM: Human steroidogenic acute regulatory protein (StAR): Functional activity in COS-1 cells, tissue-specific expression, and mapping of the structural gene to 8p11.2 and an expressed pseudogene to chromosome 13. Proc Natl Acad Sci USA 1995;92:4778–4782.

14 Papadopoulos V: Peripheral-type benzodiazepine/diazepam binding inhibitor receptor: Biological role in steroidogenic cell function. Endocr Rev 1993;14:222–240.

15 Hauet T, Yao ZX, Bose HS, Wall CT, Han Z, Li W, Hales DB, Miller WL, Culty M, Papadopoulos V: Peripheral-type benzodiazepine receptor-mediated action of steroidogenic acute regulatory protein on cholesterol entry into Leydig cell mitochondria. Mol Endocrinol 2005;19:540–554.

16 Liu J, Rone MB, Papadopoulos V: Protein-protein interactions mediate mitochondrial cholesterol transport and steroid biosynthesis. J Biol Chem 2006;281: 38879–38893.

17 Bose M, Whittal RM, Miller WL, Bose HS: Steroidogenic activity of StAR requires contact with mitochondrial VDAC1 and phosphate carrier protein. J Biol Chem 2008;283:8937–8845.

18 Bose HS, Lingappa VR, Miller WL: Rapid regulation of steroidogenesis by mitochondrial protein import. Nature 2002;417:87–91.

19 Bose HS, Whittal RM, Baldwin MA, Miller WL: The active form of the steroidogenic acute regulatory protein, StAR, appears to be a molten globule. Proc Natl Acad Sci USA 1999;96:7250–7255.

20 Baker BY, Yaworsky DC, Miller WL: A pH-dependent molten globule transition is required for activity of the steroidogenic acute regulatory protein, StAR. J Biol Chem 2005;280:41753–41760.

21 Thomas JL, Myers RP, Strickler RC: Human placental 3β-hydroxy-5-ene-steroid dehydrogenase and steroid 5 →,4-ene-isomerase: purification from mitochondria and kinetic profiles, biophysical characterization of the purified mitochondrial and microsomal enzymes. Steroid Biochem 1989;33: 209–217.

22 Luu-The V, Lechance Y, Labrie C, Leblanc G, Thomas JL, Strickler RC, Labrie F: Full length cDNA structure and deduced amino acid sequence of human 3β-hydroxy-5-ene steroid dehydrogenase. Mol Endocrinol 1989;3:1310–1312.

23 Lorence MC, Murry BA, Trant JM, Mason JI: Human 3β-hydroxysteroid dehydrogenase/$\Delta^5 \rightarrow \Delta^4$ isomerase from placenta: Expression in nonsteroidogenic cells of a protein that catalyzes the dehydrogenation/isomerization of C21 and C19 steroids. Endocrinology 1990;126:2493–2498.

24 Lee TC, Miller WL, Auchus RJ: Medroxyprogesterone acetate and dexamethasone are competitive inhibitors of different human steroidogenic enzymes. J Clin Endocrinol Metab 1999;84:2104–2110.

25 Cherradi N, Rossier MF, Vallotton MB, Timberg R, Friedberg I, Orly J, Wang XJ, Stocco DM, Capponi AM: Submitochondrial distribution of three key steroidogenic proteins (steroidogenic acute regulatory protein and cytochrome P450scc and 3β-hydroxysteroid dehydrogenase isomerase enzymes) upon stimulation by intracellular calcium in adrenal glomerulosa cells. J Biol Chem 1997;272:7899–7907.

26 Auchus RJ, Lee TC, Miller WL: Cytochrome b_5 augments the 17,20 lyase activity of human P450c17 without direct electron transfer. J Biol Chem 1998; 273:3158–3165.

27 Nakajin S, Shinoda M, Haniu M, Shively JE, Hall PF: C_{21} steroid side-chain cleavage enzyme from porcine adrenal microsomes. Purification and characterization of the 17α-hydroxylase/$C_{17,20}$ lyase cytochrome P450. J Biol Chem 1984;259:3971–3976.

28 Zuber MX, Simpson ER, Waterman MR: Expression of bovine 17α-hydroxylase cytochrome P450 cDNA in non-steroidogenic (COS-1) cells. Science 1986;234: 1258–1261.

29 Fan YS, Sasi R, Lee C, Winter JSD, Waterman MR, Lin CC: Localization of the human CYP17 gene (cytochrome P450 17α) to 10q24.3 by fluorescence in situ hybridization and simultaneous chromosome banding. Genomics 1992;14:1110–1111.

30 Picado-Leonard J, Miller WL: Cloning and sequence of the human gene encoding P450c17 (steroid 17α-hydroxylase/17,20 lyase): Similarity to the gene for P450c21. DNA 1987;6:439–448.

31 Lin D, Black SM, Nagahama Y, Miller WL: Steroid 17α-hydroxylase and 17,20 lyase activities of P450c17: Contributions of serine[106] and P450 reductase. Endocrinology 1993;132:2498–2506.

32 Yanagibashi K, Hall PF: Role of electron transport in the regulation of the lyase activity of C-21 side-chain cleavage P450 from porcine adrenal and testicular microsomes. J Biol Chem 1986;261:8429–8433.

33 Zhang L, Rodriguez H, Ohno S, Miller WL: Serine phosphorylation of human P450c17 increases 17,20 lyase activity: Implications for adrenarche and for the polycystic ovary syndrome. Proc Natl Acad Sci USA 1995;92:10619–10623.

34 Pandey AV, Mellon SH, Miller WL: Protein phosphatase 2A and phosphoprotein SET regulate androgen production by P450c17. J Biol Chem 2003;278: 2837–2844.

35 Pandey AV, Miller WL: Regulation of 17,20 lyase activity by cytochrome b_5 and by serine phosphorylation of P450c17. J Biol Chem 2005;280:13265–13271.

36 Wang M, Roberts DL, Paschke R, Shea TM, Masters BSS, Kim JJP: Three-dimensional structure of NADPH-cytochrome P450 reductase: Prototype for FMN- and FAD-containing enzymes. Proc Natl Acad Sci USA 1997;94:8411–8416.

37 Higashi Y, Yoshioka H, Yamane M, Gotoh O, Fujii-Kuriyama Y: Complete nucleotide sequence of two steroid 21-hydroxylase genes tandemly arranged in human chromosome: A pseudogene and genuine gene. Proc Natl Acad Sci USA 1986;83:2841–2845.

38 White PC, New MI, Dupont B: Structure of the human steroid 21-hydroxylase genes. Proc Natl Acad Sci USA 1986;83:5111–5115.

39 Rodrigues NR, Dunham I, Yu CY, Carroll MC, Porter RR, Campbell RD: Molecular characterization of the HLA-linked steroid 21-hydroxylase B gene from an individual with congenital adrenal hyperplasia. EMBO J 1987;6:1653–1661.

40 Casey ML, MacDonald PC: Extra-adrenal formation of a mineralocorticoid: Deoxycorticosterone and deoxycorticosterone sulfate biosynthesis and metabolism. Endocr Rev 1982;3:396–403.

41 Mellon SH, Miller WL: Extra-adrenal steroid 21-hydroxylation is not mediated by P450c21. J Clin Invest 1989;84:1497–1502.

42 Yamazaki H, Shimada T: Progesterone and testosterone hydroxylation by cytochromes P450 2C19, 2C9 and 3A4. Arch Biochem Biophys 1997;346:161–169.

43 White PC, Curnow KM, Pascoe L: Disorders of steroid 11β-hydroxylase isozymes. Endocr Rev 1994; 15:421–438.

44 Pascoe L, Curnow K, Slutsker L, Rösler A, White PC: Mutations in the human CYP11B2 (aldosterone synthase) gene causing corticosterone methlyoxidase II deficiency. Proc Natl Acad Sci USA 1992;89: 4996–5000.

45 Ulick S, Wang JZ, Morton H: The biochemical phenotypes of two inborn errors in the biosynthesis of aldosterone. J Clin Endocrinol Metab 1992;74:1415–1420.

46 Zhang G, Rodriguez H, Fardella CE, Harris DA, Miller WL: Mutation T318M in P450c11AS causes corticosterone methyl oxidase II deficiency. Am J Hum Genet 1995;57:1037–1043.

47 Moghrabi N, Andersson S: 17β-hydroxysteroid dehydrogenases: Physiological roles in health and disease. Trends Endocrinol Metab 1998;9:265–270.

48 Penning TM: Molecular endocrinology of hydroxysteroid dehydrogenases. Endocr Rev 1997;18:281–305.

49 Peltoketo H, Isomaa V, Mäenlavsta O, Vihko R: Complete amino acid sequence of human placental 17β-hydroxysteroid dehydrogenase deduced from cDNA. FEBS Lett 1988;239:73–77.

50 Tremblay Y, Ringler GE, Morel Y, Mohandas TK, Labrie F, Strauss JF III, Miller WL: Regulation of the gene for estrogenic 17-ketosteroid reductase lying on chromosome 17cen→q25. J Biol Chem 1989;264: 20458–20462.

51 Ghosh D, Pleuteu VZ, Zhu DW, Wawarzak Z, Duax WL, Plaugborn W, Labrie F, Lin S-X: Structure of human estrogenic 17β-hydroxysteroid dehydrogenase at 2.2 Å resolution. Structure 1995;3:503–513.

52 Takeyama J, Sasano H, Suzuki T, Iinuma K, Nagura H, Andersson S: 17β-hydroxysteroid dehydrogenase types 1 and 2 in human placenta: an immunohistochemical study with correlation to placental development. J Clin Endocrinol Metab 1998;83: 3710–3715.

53 Geissler WM, Davis DL, Wu L, Bradshaw KD, Patel S, Mendonça BB, Elliston KO, Wilson JD, Russell DW, Andersson S: Male pseudohermaphroiditism caused by mutations of testicular 17β-hydroxysteroid dehydrogenase 3. Nat Genet 1994;7:34–39.

54 Adamski J, Normand T, Leenders F, Montae D, Begue A, Staehelin D, Jungblut PW, de Launoit Y: Molecular cloning of a novel widely expressed human 80 kDa 17 β-hydroxysteroid dehydrogenase IV. Biochem J 1995;311:437–443.

55 Leenders F, Tesdorpf JG, Markus M, Engel T, Seedorf U, Adamski J: Porcine 80-kDa protein reveals intrinsic 17 β-hydroxysteroid dehydrogenase, fatty acyl-CoA-hydratase/dehydrogenase, and sterol transfer activities. J Biol Chem 1996;271:5438–5442.

56 van Grunsven EG, van Berkel E, Ijlst L, Vreken P, De Klerk JBC, Adamski J, Lemonde H, Clayton PT, Cuebas DA, Wanders RJA: Peroxisomal D-hydroxyacyl-CoA dehydrogenase deficiency: Resolution of the enzyme defect and its molecular basis in bifunctional protein deficiency. Proc Natl Acad Sci USA 1998;95:2128–2133.

57 Lin HK, Jez JM, Schlegel BP, Peehl DM, Pachter JA, Penning TM: Expression and characterization of recombinant type 2 3α-hydroxysteroid dehydrogenase (HSD) from human prostate: demonstration of bifunctional 3α/17β-HSD activity and cellular distribution. Mol Endocrinol 1997;11:1971–1984.

58 Dufort I, Rheault P, Huang XF, Soucy P, Luu-The V: Characteristics of a highly labile human type 5 17β-hydroxysteroid dehydrogenase. Endocrinology 1999; 140:568–574.

59 Falany CN: Enzymology of human cystolic sulfotransferases. FASEB J 1997;11:206–216.

60 Stroh CA: Sulfonation and molecular action. Endocr Rev 2002;23:703–732.

61 Tong MH, Jiang H, Liu P, Lawson JA, Brass LF, Song W-C: Spontaneous fetal loss caused by placental thrombosis in estrogen sulfutransferase-deficient mice. Nat Med 2005;11:153–159.

62 Nowell S, Falany CN: Pharmacogenetics of human cytosolic sulfotransferases. Oncogene 2006;25: 1673–1678.

63 Yen PH, Allen E, Marsh B, Mohandas T, Shapiro LJ: Cloning and expression of steroid sulfatase cDNA and the frequent occurrence of deletions: Implications for X-Y interchange. Cell 1987;49:443–454.

64 Compagnone NA, Salido E, Shapiro LJ, Mellon SH: Expression of steroid sulfatase during embryogenesis. Endocrinology 1997;138:4768–4773.

65 Simpson ER, Mahendroo MS, Means GD, Kilgore MW, Hinshelwood MM, Graham-Lorence S, Amarneh B, Ito Y, Fisher CR, Michael MD, Mendelson CR, Bulun SE: Aromatase cytochrome P450, the enzyme responsible for estrogen biosynthesis. Endocr Rev 1994;15:342–355.

66 Grumbach MM, Auchus RJ: Estrogen: consequences and implications of human mutations in synthesis and action. J Clin Endocrinol Metab 1999; 84:4677–4694.

67 Conte FA, Grumbach MM, Ito Y, Fisher CR, Simpson ER: A syndrome of female pseudohermaphrodism, hypergonadotropic hypogonadism, and multicystic ovaries associated with missense mutations in the gene encoding aromatase (P450arom). J Clin Endocrinol Metab 1994;78:1287–1292.

68 Thigpen AE, Silver RI, Guileyardo JM, Casey ML, McConnell JD, Russell DW: Tissue distribution and ontogeny of steroid 5α-reductase isozyme expression. J Clin Invest 1993;92:903–910.

69 Wilson JD: The role of androgens in male gender role behavior. Endocr Rev 1999;20:726–737.

70 White PC, Mune T, Agarwal AK: 11β-hydroxysteroid dehydrogenase and the syndrome of apparent mineralocorticoid excess. Endocr Rev 1997;18:135–156.

71 Agarwal AK, Tusie-Luna M-T, Monder C, White PC: Expression of 11β-hydroxysteroid dehydrogenase using recombinant vaccinia virus. Mol Endocrinol 1990;4:1827–1832.

72 Moore CCD, Mellon SH, Murai J, Siiteri PK, Miller WL: Structure and function of the hepatic form of 11β-hydroxysteroid dehydrogenase in the squirrel monkey, an animal model of glucocorticoid resistance. Endocrinology 1993;133:368–375.

73 Rusvai E, Náray-Fejes-Tóth A: A new isoform of 11β-hydroxysteroid dehydrogenase in aldosterone target cells. J Biol Chem 1993;268:10717–10720.

74 Hirasawa G, Sasono H, Suzuki T, Takeyama J, Muramatu Y, Fukushima K, Hiwatashi N, Toyota T, Nagura H, Krozowski Z: 11β-hydroxysteroid dehydrogenase type 2 and mineralocorticoid receptor in human fetal development. J Clin Endocrinol Metab 1999;84:1453–1458.

75 Hewitt KN, Walker EA, Stewart PM: Hexose-6-phosphate dehydrogenase and redox control of 11β-hydroxysteroid dehydrogenase type 1 activity. Endocrinology 2005;146:2539–2543.

76 Goto M, Hanley KP, Marcos J, Wood PJ, Wright S, Postle AD, Cameron IT, Mason JI, Wilson DI, Hanley NA: In humans, early cortisol biosynthesis provides a mechanism to safeguard female sexual development. J Clin Invest 2006;116:953–960.

77 Fujieda K, Faiman C, Feyes FI, Winter JSD: The control of steroidogenesis by human fetal adrenal cells in tissue culture: IV. The effects of exposure to placental steroids. J Clin Endocrinol Metab 1982; 54:89–94.

78 DiBlasio AM, Voutilainen R, Jaffe RB, Miller WL: Hormonal regulation of mRNAs for P450scc (cholesterol side-chain cleavage enzyme) and P450c17 (17α-hydroxylase/17,20 lyase) in cultured human fetal adrenal cells. J Clin Endocrinol Metab 1987;65: 170–175.

79 New MI, Wilson RC: Steroid disorders in children: congenital adrenal hyperplasia and apparent mineralocorticoid excess. Proc Natl Acad Sci USA 1999; 96:12790–12797.

Walter L. Miller, MD
Department of Pediatrics, Division of Endocrinology, Bldg. 672-S
University of California
San Francisco, CA 94142-0978 (USA)
Tel. +1 415 476 2598, Fax +1 415 476 6286, E-Mail wlmlab@itsa.ucsf.edu

Flück CE, Miller WL (eds): Disorders of the Human Adrenal Cortex.
Endocr Dev. Basel, Karger, 2008, vol 13, pp 19–32

Disorders of Adrenal Development

Bruno Ferraz-de-Souza · John C. Achermann

Developmental Endocrinology Research Group, Clinical and Molecular Genetics Unit,
UCL Institute of Child Health, University College London, London, UK

Abstract

Human adrenal development is a complex and relatively poorly understood process. However, significant insight into some of the mechanisms regulating adrenal development and function is being obtained through the analysis of individuals and families with adrenal hypoplasia. Adrenal hypoplasia can occur: (1) secondary to defects in pituitary adrenocorticotropin (ACTH) synthesis, processing and release (secondary adrenal hypoplasia; e.g. HESX1, LHX4, SOX3, TPIT, pituitary POMC, PC1); (2) as part of several ACTH resistance syndromes (e.g. MC2R/ACTHR, MRAP, Alacrima, Achalasia, Addison disease), or as (3) a primary defect in the development of the adrenal gland itself (primary adrenal hypoplasia; e.g. DAX1/*NR0B1* – dosage-sensitive sex reversal, adrenal hypoplasia congenita critical region on the X chromosome 1). Indeed, the X-linked form of primary adrenal hypoplasia due to deletions or mutations in the orphan nuclear receptor DAX1 occurs in around half of male infants presenting with a salt-losing adrenal crisis, where no obvious steroidogenic defect (e.g. 21-hydroxylase deficiency), metabolic abnormality (e.g. neonatal adrenoleukodystrophy) or physical cause (e.g. adrenal haemorrhage) is found. Establishing the underlying basis of adrenal failure can have important implications for investigating associated features, the likely long-term approach to treatment, and for counselling families about the risk of other children being affected.

Disorders of adrenal development generally result in small, poorly functioning glands: a clinical condition termed 'adrenal hypoplasia' [1, 2]. Although the complex mechanisms underlying adrenal development in humans remain relatively poorly understood, significant insight into some of the factors involved in adrenal development and function is being obtained from studies of families with these conditions. Several single gene disorders have been identified in recent years that can affect the hypothalamic-pituitary adrenal (HPA) axis at different levels, and a genetic cause is found in approximately 50% of all individuals with secondary or primary forms of adrenal hypoplasia. This chapter will provide a brief overview of some current concepts of adrenal development in humans, and will describe several of the more important genetic causes of adrenal hypoplasia that can present with adrenal dysfunction in infancy, childhood or even adulthood.

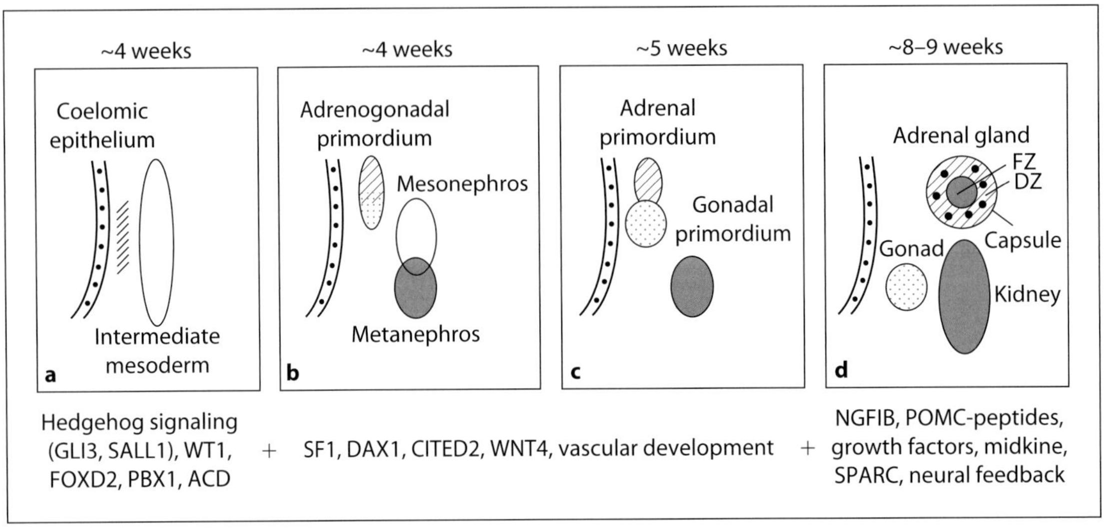

Fig. 1. Cartoon showing key events in early human adrenal development. FZ = Fetal zone; DZ = definitive zone. Reproduced with permission from Flück et al. [50]. Copyright Elsevier, 2008.

Normal Adrenal Development

Embryology

The adrenal cortex is derived from a thickening of the intermediate mesoderm, which occurs at around 4–5 weeks of gestation in humans [1, 3] (fig. 1). This region contains adrenogonadal progenitor cells that give rise to the steroidogenic cells of the adrenal gland as well as those of the gonad. Cells destined to become adrenal tissue migrate retroperitoneally to the upper pole of the mesonephros and are infiltrated at around 7–8 weeks' gestation by sympathetic cells derived from the neural crest that give rise to the adrenal medulla. Encapsulation of the adrenal gland occurs sometime after 8 weeks' gestation, resulting in the formation of a distinct organ just above the developing kidney.

Adrenal Zonation and Growth

The developing fetal adrenal cortex consists of an outer 'definitive' zone that can synthesize glucocorticoids and mineralocorticoids, and a much larger inner 'fetal' zone that produces significant amounts of androgenic precursors (e.g. dehydroepiandrosterone – DHEA, and dehydroepiandrosterone sulphate – DHEAS), which are converted to oestrogens by the placenta [4]. The fetal zone is a characteristic feature of higher primates, although the biological role of these fetal androgens – if any – is unclear. Nevertheless, the fetal zone enlarges rapidly throughout pregnancy so that the developing adrenal glands are huge structures that weight roughly the same at birth as during adulthood, and represent 0.4% of body weight at term [4, 5]. The adrenal gland undergoes rapid involution post-natally. This process is largely due to regression of the fetal zone, which is absent by 6 months of age in most cases. A putative 'transitional' zone has been described between the definitive and fetal zones during later fetal development, but its role is unclear. This region was once proposed to contain stem cells that could differentiate into either definitive type or fetal type tissue, although other hypotheses propose that the main population of adrenal stem cells is located in the subcapsular region of the gland, and that cells mature through different stages of development as they migrate in a 'centripetal' manner [4].

Several factors have been identified that play potentially important roles in regulating adrenal development, zonation and growth. Many of these factors have been found following studies of transgenic mice and of patients with various forms of adrenal hypoplasia, although the interaction and relative significance of many of these factors is currently poorly understood. For example, the earliest stages of adrenal development may be regulated by a number of transcription factors [e.g. SAL-like 1 – SALL1, human forkhead-box gene D2 – FOXD2, pre-B-cell leukaemia transcription factor 1 – PBX1, Wilms tumour 1 gene – WT1, steroidogenic factor 1 – SF1 (*NR5A1*), DAX1 (*NR0B1*)], co-regulators (e.g. CBP/p300-interacting transactivator, with Glu/Asp-rich C-terminal domain 2 – CITED2), signalling factors (e.g. hedgehog/GLI-Kruppel family member 3 – GLI3, wingless type MMTV integration site family members WNT3/WNT4/WNT11, midkine), matrix proteins (e.g. secreted protein, acidic, cysteine-rich – SPARC) and regulators of telomerase activity (e.g. adrenocortical dysplasia – ACD; fig. 1) [1, 3, 6–11]. Subsequently, fetal adrenal growth is highly-dependent upon the tropic effects of adrenocorticotropic hormone (ACTH) stimulation and other growth factor signalling pathways such as basic fibroblast growth factor, epidermal growth factor and insulin-like growth factor II.

Fetal Adrenal Steroidogenesis

The capacity of the fetal zone of the adrenal gland to produce large amounts of androgenic precursors (DHEA, DHEAS) is largely the result of a relative deficiency of 3β-hydroxysteroid dehydrogenase type II activity coupled with a relative abundance of the 17,20-lyase activity of P450c17 [4]. The fetal adrenal also has significant expression of sulphotransferase, which drives the conversion of DHEA to DHEAS and provides a substrate for conversion to circulating oestrogens by the placenta.

The definitive zone of the fetal adrenal gland is also active in early gestation, and is able to produce glucocorticoids in the first trimester due to the transient expression of 3β-hydroxysteroid dehydrogenase type II, which is upregulated between 8 and 9 weeks' gestation [5]. Recent studies have shown that the HPA axis is sensitive to glucocorticoid-mediated feedback during this time; thus, 46,XX fetuses with steroidogenic defects such as 21-hydroxylase or 11-hydroxylase deficiencies lack cortisol and have an elevated ACTH drive that results in excess production of fetal androgens at a time when

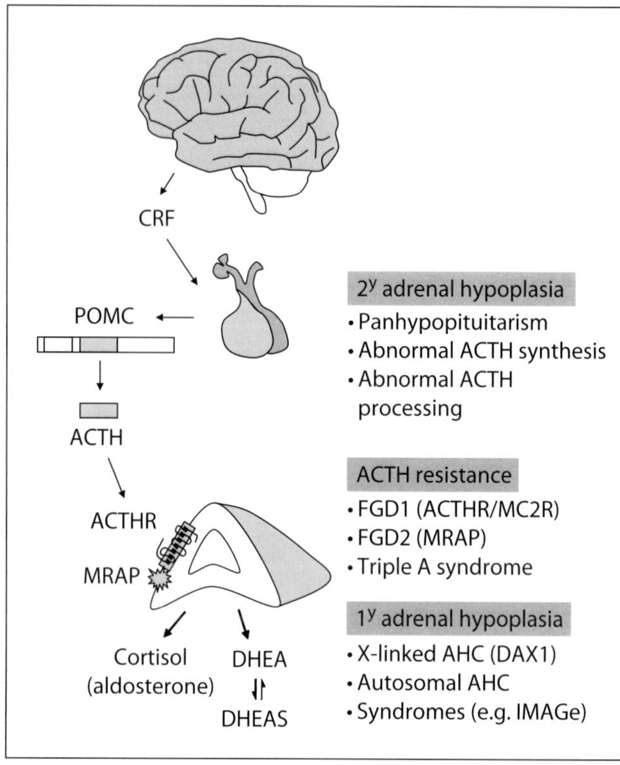

Fig. 2. Overview of the HPA axis showing the different types of adrenal hypoplasia. FGD = Familial glucocorticoid deficiency.

the genital tubercle and scrotal folds are sensitive to androgen exposure [5]. This imbalance results in androgenisation of the female fetus' genitalia. Following this period of transient intact HPA axis activity, fetal adrenal glucocorticoid production is reduced as 3β-hydroxysteroid dehydrogenase type II activity declines. However, glucocorticoid production resumes in the third trimester so that the fetus is primed for post-natal existence, and development of the zona glomerulosa means that the adrenal is capable of responding to angiotensin II by producing mineralocorticoids after birth.

Disorders of Adrenal Development

Significant insight into our understanding of adrenal development has been obtained following the identification of individuals and families with inherited disorders of adrenal development in recent years. These conditions can be subdivided into: (1) secondary adrenal hypoplasia due to defects in ACTH synthesis, processing and release; (2) ACTH resistance syndromes, and (3) primary adrenal hypoplasia due to defects in the development of the adrenal gland itself (fig. 2; table 1).

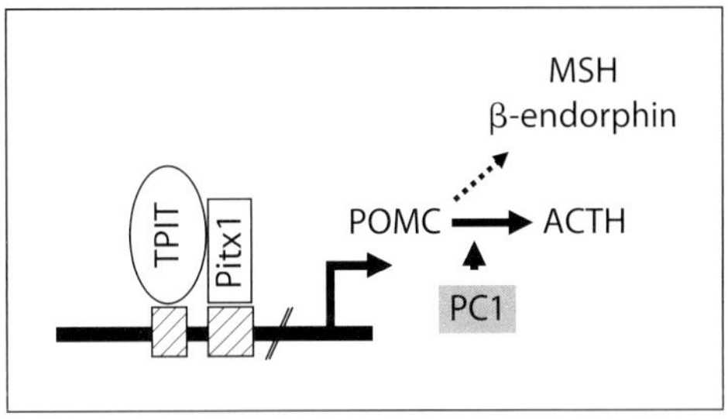

Fig. 3. Diagrammatic representation of the processes involved in POMC synthesis and cleavage in the corticotrope. Modified with permission from Lin and Achermann [2]. Copyright Blackwell, 2004.

role in appetite regulation and weight, as well as pigmentation of the skin and hair. Thus, defects that involve POMC itself have more widespread consequences and are associated with pale skin, red hair and obesity in addition to ACTH deficiency [16]. These cutaneous features may be less marked in individuals with dark hair, and may diminish with age [17]. Finally, processing of POMC into the mature ACTH peptide requires the actions of the cleavage enzyme prohormone convertase-1 (PC1, also known as proprotein convertase, subtilisin/kexin-type, 1/*PCSK1*). Abnormalities in ACTH processing due to defects in PC1 can cause secondary adrenal failure in rare cases [18]. As the processing of several other peptide hormones is disrupted, associated features include obesity, hypogonadism, hypoglycaemia and persistent malabsorptive diarrhoea [19].

Adrenocorticotropin Resistance Syndromes

ACTH resistance can occur in a number of well-defined conditions, such as defects in the ACTH receptor (melanocortin receptor 2 – MC2R, familial glucocorticoid deficiency type 1); MC2R accessory protein (MRAP, familial glucocorticoid deficiency type 2); or as part of the triple A syndrome (Alacrima, Achalasia, Addison; also known as Allgrove syndrome and due to defects in ALADIN/AAAS) [20]. In general, these conditions present with isolated glucocorticoid deficiency, hyperpigmentation and markedly elevated ACTH. Approximately 15% of individuals with triple A syndrome have evidence of mineralocorticoid insufficiency, and children with severe disruptive changes in the ACTH receptor may have evidence of mild hyponatraemia at presentation [21]. A complete review of ACTH resistance syndromes and associated features is provided in the chapter by Cooray et al. [pp. 99–116].

Primary Adrenal Hypoplasia

Adrenal hypoplasia congenita (AHC), also known as congenital adrenal hypoplasia, is a disorder of adrenal development resulting in primary adrenal insufficiency. This condition usually presents with severe salt-losing primary adrenal failure in early infancy or childhood, although milder, delayed onset forms of the condition exist. The inheritance pattern or associated or syndromic features might help to point to a specific diagnosis.

X-Linked Adrenal Hypoplasia

X-linked AHC results from mutations in the nuclear receptor DAX1 *(NR0B1)*. This condition is the most prevalent form of primary adrenal hypoplasia reported to date [22, 23]. X-linked AHC was probably first described in 1948 in an infant who died at 33 days of age with hyperpigmentation and small adrenal glands. The presence of some 'cytomegalic' cells typical of fetal zone adrenal tissue led to this condition being termed 'cytomegalic adrenal hypoplasia'. The X-linked pattern of inheritance of AHC became apparent in the 1960s and an association with hypogonadotropic hypogonadism was described as boys who received steroid treatment did not progress through puberty.

The gene for X-linked AHC was found to be located on the short arm of the X-chromosome (Xp21.3) and was identified as DAX1 *(NR0B1)* in 1994 [24, 25]. This discovery was helped by many reports of X-linked AHC as part of a contiguous gene deletion syndrome. Associated loci include those for glycerol kinase deficiency, ornithine transcarbamylase deficiency and Duchenne muscular dystrophy centromeric to DAX1 *(NR0B1)*, as well as the telomeric gene *IL1RAPL1* associated with an X-linked form of developmental delay, which is deleted in a small number of cases.

DAX1 is a member of the 'orphan' nuclear receptor superfamily. The carboxyl terminal region of DAX1 has sequence homology to the ligand-binding domain of nuclear receptors but no naturally-occurring ligand for DAX1 has been identified (fig. 4). The amino-terminus of DAX1 is believed to contain an unusual repeat motif structure that contains several LXXLL domains involved in nuclear receptor/co-factor interaction. DAX1 is expressed in the developing adrenal gland, gonad and gonadotropes in keeping with its important role in the development and regulation of these structures.

X-linked AHC due to DAX1 mutations is characterized by: (1) primary adrenal insufficiency; (2) hypogonadotropic hypogonadism, and (3) a likely primary defect in spermatogenesis. Boys tend to present with salt-losing adrenal failure in the first 2 months of life (60–70%) or more insidiously with adrenal failure throughout childhood (30–40%) [23, 26]. Isolated mineralocorticoid deficiency may be the presenting feature in some cases and cortisol levels may appear normal initially; however, glucocorticoid deficiency usually develops with time [27, 28]. Absent or arrested puberty due to a combined hypothalamic and pituitary defect typically occurs during adolescence [29]. However, several reports of limited testicular enlargement or signs of premature sexual maturation in childhood have been published [30, 31]. Generation of an *Ahch (Dax1)*-deleted mouse using a Cre-recombinase strategy has shown an intrinsic spermatogenic defect associated with DAX1 deficiency [32]. The extent of

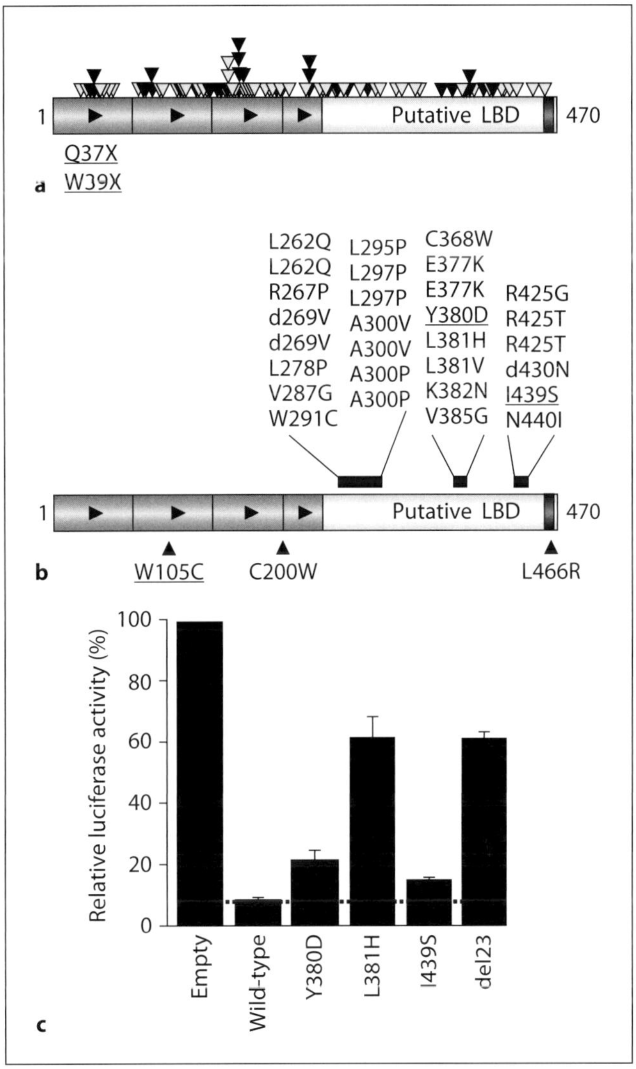

Fig. 4. DAX1 *(NR0B1)* is an orphan nuclear receptor with an atypical amino-terminal repeat motif structure and a carboxyl terminal region that resembles a ligand-binding domain. **a** A selection of frameshift (grey arrowheads) and nonsense (black arrowheads) mutations found in individuals with X-linked AHC. **b** Missense mutations in DAX1 tend to cluster within certain regions of the carboxyl terminus of DAX1 (black bars). Those changes associated with a variant or late-onset phenotypes are underlined. **c** Functional assay of DAX1 as a repressor of gene transcription. Wild-type (WT) DAX1 represses luciferase activity in this in vitro assay compared to empty vector. Point mutations associated with a classic X-linked AHC phenotype (L381H, del23 amino-acids from the carboxyl-terminus) cause loss of function, whereas those changes associated with a delayed-onset adrenal failure (Y380D, I439S) have partial loss of function. Modified with permission from Lin et al. [23] and from Mantovani et al. [38]. Copyright The Endocrine Society, 2006 and 2002.

the spermatogenic defect in humans is still unclear, but spontaneous fertility is extremely rare in men with X-linked AHC, and the results of fertility induction using gonadotropins have been disappointing to date [33]. It is not yet known whether techniques such as intracytoplasmic sperm injection will be successful.

Initial case reports of X-linked AHC were understandably biased to reporting individuals with contiguous gene deletion syndromes. However, the identification of DAX1 as the gene responsible for this condition has allowed increasing insight into the relative prevalence of X-linked AHC as well as helping to elucidate some of the pathogenic mechanisms of this condition. More than 100 different DAX1 mutations have been reported to date in more than 200 individuals and families with X-linked AHC (fig. 4) [22, 23]. An analysis of 37 cases of X-linked AHC from a single centre over the past 10 years has shown isolated DAX1 gene deletions in 8 (22%) cases, contiguous gene deletions in 2 (5%) cases, and point mutations in the rest [nonsense, 7 (19%); frameshift, 12 (32%); missense, 8 (22%)] (fig. 4) [23]. Missense mutations tend to cluster within certain regions of the ligand-like binding domain, but rare amino terminal missense mutations have been described (fig. 4) [28, 34]. These point changes may interfere with nuclear localization as well as affecting protein-protein interaction [35]. Nonsense and frameshift mutations are located throughout the DAX1 gene and loss of the carboxyl terminal region of the protein (contributing to the activation function 2 domain) is sufficient for complete loss of protein function in most cases (fig. 4).

In this analysis of 64 boys with AHC, DAX1 mutations were found in all individuals with primary adrenal failure, abnormal puberty and a family history of adrenal disease in males (8/8, 100%), but also in approximately 40% of a cohort of prepubertal boys with no family history of note, in whom other diagnoses such as congenital adrenal hyperplasia (e.g. 21-hydroxylase deficiency) and metabolic defects (e.g. adrenoleukodystrophy) had been excluded [23]. Genetic analysis of DAX1 for individuals with X-linked AHC is now available as a clinical test, and making the genetic diagnosis of a DAX1 mutation has significant implications for planning future management and for predicting the need to induce puberty (e.g. see www.genetests.org). As this is an X-linked condition, close monitoring and genetic counselling can help to prevent life-threatening adrenal crises in other family members or future pregnancies [36].

In addition to the phenotypes described above, several atypical phenotypes have now been described in association with DAX1 mutations. A delayed-onset form of X-linked AHC has been described in men who presented between 20–30 years of age with mild primary adrenal insufficiency or partial hypogonadism. Some of these patients harbour missense mutations (I439S, Y380D) that have limited DAX1 function (fig. 4) [37, 38]. In other individuals, nonsense mutations at the extreme amino-terminal region of DAX-1 may be associated with the translation of an alternate in-frame DAX1 isoform from a methionine residue at codon 83 [39]. This amino-terminally truncated protein retains an LXXLL domain and has partial activity, consistent with the milder phenotype seen in the patient. Finally, skewed X-inactivation may result in delayed puberty or even primary adrenal failure in girls or women who have heterozygous DAX1 changes [40].

Despite the significant number of DAX1 mutations described, the exact molecular pathogenesis of X-linked AHC remains unclear. To date, most in vitro functional studies have suggested that DAX1 acts as a repressor of transcription, with a putative interaction with the related nuclear receptor steroidogenic factor-1 (SF1) [34, 41]. Other hypotheses propose that DAX1 may regulate adrenal progenitor cell development and maturation so that loss of DAX1 function is associated with aberrant cellular differentiation [42].

Autosomal Adrenal Hypoplasia

The underlying basis of autosomal forms of adrenal hypoplasia remains poorly understood. Heterozygous or homozygous mutations in the nuclear receptor steroidogenic factor-1 (SF1, *NR5A1*) have been reported in 46,XY phenotypic females with either spontaneous or recessively inherited primary adrenal failure, and a heterozygous SF1 mutation has been described in a 46,XX girl with adrenal dysfunction [43–45]. However, SF1 mutations have not been found in phenotypic males with adrenal hypoplasia [23]. It is likely that other autosomal genes are responsible for these rare recessive forms of adrenal hypoplasia. Furthermore, defects in P450 side chain cleavage enzyme (CYP11A1) can cause salt-losing adrenal failure with small adrenals on imaging [46] and partial defects in steroidogenic acute regulatory protein (StAR) may mimic ACTH resistance [47].

Syndromic Forms of Adrenal Hypoplasia

Primary adrenal failure can sometimes occur in conditions such as Pena-Shokeir syndrome type I, pseudotrisomy 13, and Meckel syndrome [1]. Primary adrenal hypoplasia is also an important feature of IMAGe syndrome (intrauterine growth retardation, metaphyseal dysplasia, adrenal hypoplasia, genitourinary anomalies) [48]. The underlying aetiology of this condition remains unknown [49]. It is hoped that a better understanding of human adrenal development, analysis of pedigrees where adrenal hypoplasia is a feature, and more detailed analysis of transgenic mouse phenotypes will provide more candidate genes for syndromes such as IMAGe, as well as for those children and adults with adrenal hypoplasia in whom the cause is currently unknown. It can be viewed as a great success that the molecular aetiology of developmental adrenal disorders can be found in over 50% of children to date; the next decade will hopefully reveal several other significant causes of adrenal hypoplasia in those individuals and families where the underlying pathogenesis currently remains elusive.

Acknowledgements

B.F.S. holds a scholarship from Capes/Brazil (4798066). J.C.A. holds a Wellcome Trust Senior Research Fellowship in Clinical Science (079666).

References

1 Else T, Hammer GD: Genetic analysis of adrenal absence: agenesis and aplasia. Trends Endocrinol Metab 2005;16:458–468.

2 Lin L, Achermann JC: Inherited adrenal hypoplasia: not just for kids! Clin Endocrinol (Oxf) 2004;60: 529–537.

3 Hammer GD, Parker KL, Schimmer BP: Minireview: transcriptional regulation of adrenocortical development. Endocrinology 2005;146:1018–1024.

4 Mesiano S, Jaffe RB: Developmental and functional biology of the primate fetal adrenal cortex. Endocr Rev 1997;18:378–403.

5 Goto M, Piper HK, Marcos J, Wood PJ, Wright S, Postle AD, Cameron IT, Mason JI, Wilson DI, Hanley NA: In humans, early cortisol biosynthesis provides a mechanism to safeguard female sexual development. J Clin Invest 2006;116:953–960.

6 Schnabel CA, Selleri L, Cleary ML: Pbx1 is essential for adrenal development and urogenital differentiation. Genesis 2003;37:123–130.

7 Luo X, Ikeda Y, Parker KL: A cell-specific nuclear receptor is essential for adrenal and gonadal development and sexual differentiation. Cell 1994;77: 481–490.

8 Bamforth SD, Braganca J, Eloranta JJ, Murdoch JN, Marques FI, Kranc KR, Farza H, Henderson DJ, Hurst HC, Bhattacharya S: Cardiac malformations, adrenal agenesis, neural crest defects and exencephaly in mice lacking Cited2, a new Tfap2 co-activator. Nat Genet 2001;29:469–474.

9 Dewing P, Ching ST, Zhang YH, Huang BL, Peirce RM, McCabe ER, Vilain E: Midkine is expressed early in rat fetal adrenal development. Mol Genet Metab 2000;71:616–622.

10 Ishimoto H, Ginzinger DG, Matsumoto T, Hattori Y, Furuya M, Minegishi K, Tanaka M, Yoshimura Y, Jaffe RB: Differential zonal expression and adrenocorticotropin regulation of secreted protein acidic and rich in cysteine (SPARC), a matricellular protein, in the midgestation human fetal adrenal gland: implications for adrenal development. J Clin Endocrinol Metab 2006;91:3208–3214.

11 Keegan CE, Hutz JE, Else T, Adamska M, Shah SP, Kent AE, Howes JM, Beamer WG, Hammer GD: Urogenital and caudal dysgenesis in adrenocortical dysplasia (acd) mice is caused by a splicing mutation in a novel telomeric regulator. Hum Mol Genet 2005;14:113–123.

12 Kelberman D, Dattani MT: The role of transcription factors implicated in anterior pituitary development in the aetiology of congenital hypopituitarism. Ann Med 2006;38:560–577.

13 Pernasetti F, Toledo SP, Vasilyev VV, Hayashida CY, Cogan JD, Ferrari C, Lourenco DM, Mellon PL: Impaired adrenocorticotropin-adrenal axis in combined pituitary hormone deficiency caused by a two-base pair deletion (301–302delAG) in the prophet of Pit-1 gene. J Clin Endocrinol Metab 2000; 85:390–397.

14 Pulichino AM, Vallette-Kasic S, Couture C, Gauthier Y, Brue T, David M, Malpuech G, Deal C, Van VG, De VM, Riepe FG, Partsch CJ, Sippell WG, Berberoglu M, Atasay B, Drouin J: Human and mouse TPIT gene mutations cause early onset pituitary ACTH deficiency. Genes Dev 2003;17:711–716.

15 Vallette-Kasic S, Brue T, Pulichino AM, Gueydan M, Barlier A, David M, Nicolino M, Malpuech G, Dechelotte P, Deal C, Van VG, De VM, Riepe FG, Partsch CJ, Sippell WG, Berberoglu M, Atasay B, de ZF, Beckers D, Kyllo J, Donohoue P, Fassnacht M, Hahner S, Allolio B, Noordam C, Dunkel L, Hero M, Pigeon B, Weill J, Yigit S, Brauner R, Heinrich JJ, Cummings E, Riddell C, Enjalbert A, Drouin J: Congenital isolated adrenocorticotropin deficiency: an underestimated cause of neonatal death, explained by TPIT gene mutations. J Clin Endocrinol Metab 2005; 90:1323–1331.

16 Krude H, Biebermann H, Luck W, Horn R, Brabant G, Gruters A: Severe early-onset obesity, adrenal insufficiency and red hair pigmentation caused by POMC mutations in humans. Nat Genet 1998;19: 155–157.

17 Krude H, Biebermann H, Schnabel D, Tansek MZ, Theunissen P, Mullis PE, Gruters A: Obesity due to proopiomelanocortin deficiency: three new cases and treatment trials with thyroid hormone and ACT H4–10. J Clin Endocrinol Metab 2003;88:4633–4640.

18 Jackson RS, Creemers JW, Ohagi S, Raffin-Sanson ML, Sanders L, Montague CT, Hutton JC, O'Rahilly S: Obesity and impaired prohormone processing associated with mutations in the human prohormone convertase 1 gene. Nat Genet 1997;16:303–306.

19 Jackson RS, Creemers JW, Farooqi IS, Raffin-Sanson ML, Varro A, Dockray GJ, Holst JJ, Brubaker PL, Corvol P, Polonsky KS, Ostrega D, Becker KL, Bertagna X, Hutton JC, White A, Dattani MT, Hussain K, Middleton SJ, Nicole TM, Milla PJ, Lindley KJ, O'Rahilly S: Small-intestinal dysfunction accompanies the complex endocrinopathy of human proprotein convertase 1 deficiency. J Clin Invest 2003;112: 1550–1560.

20 Metherell LA, Chan LF, Clark AJ: The genetics of ACTH resistance syndromes. Best Pract Res Clin Endocrinol Metab 2006;20:547–560.

21 Lin L, Hindmarsh PC, Metherell LA, Alzyoud M, Al-Ali M, Brain CE, Clark AJ, Dattani MT, Achermann JC: Severe loss-of-function mutations in the adrenocorticotropin receptor (ACTHR, MC2R) can be found in patients diagnosed with salt-losing adrenal hypoplasia. Clin Endocrinol (Oxf) 2007;66: 205–210.

22 Phelan JK, McCabe ER: Mutations in NR0B1 (DAX1) and NR5A1 (SF1) responsible for adrenal hypoplasia congenita. Hum Mutat 2001;18:472–487.

23 Lin L, Gu WX, Ozisik G, To WS, Owen CJ, Jameson JL, Achermann JC: Analysis of DAX1 (NR0B1) and steroidogenic factor-1 (NR5A1) in children and adults with primary adrenal failure: ten years' experience. J Clin Endocrinol Metab 2006;91:3048–3054.

24 Muscatelli F, Strom TM, Walker AP, Zanaria E, Recan D, Meindl A, Bardoni B, Guioli S, Zehetner G, Rabl W, Schwarz HP, Kaplan J-C, Camerino G, Meitinger T, Monaco AP: Mutations in the DAX-1 gene give rise to both X-linked adrenal hypoplasia congenita and hypogonadotropic hypogonadism. Nature 1994;372:672–676.
25 Zanaria E, Muscatelli F, Bardoni B, Strom TM, Guioli S, Guo W, Lalli E, Moser C, Walker AP, McCabe ER, Meitinger T, Monaco AP, Sassone-Corsi P, Camerino G: An unusual member of the nuclear hormone receptor superfamily responsible for X-linked adrenal hypoplasia congenita. Nature 1994;372:635–641.
26 Reutens AT, Achermann JC, Ito M, Ito M, Gu WX, Habiby RL, Donohoue PA, Pang S, Hindmarsh PC, Jameson JL: Clinical and functional effects of mutations in the DAX-1 gene in patients with adrenal hypoplasia congenita. J Clin Endocrinol Metab 1999; 84:504–511.
27 Wiltshire E, Couper J, Rodda C, Jameson JL, Achermann JC: Variable presentation of X-linked adrenal hypoplasia congenita. J Pediatr Endocrinol Metab 2001;14:1093–1096.
28 Verrijn Stuart AA, Ozisik G, de Vroede MA, Giltay JC, Sinke RJ, Peterson TJ, Harris RM, Weiss J, Jameson JL: An amino-terminal DAX1 (NROB1) missense mutation associated with isolated mineralocorticoid deficiency. J Clin Endocrinol Metab 2007;92:755–761.
29 Habiby RL, Boepple P, Nachtigall L, Sluss PM, Crowley WF Jr, Jameson JL: Adrenal hypoplasia congenita with hypogonadotropic hypogonadism: evidence that DAX-1 mutations lead to combined hypothalmic and pituitary defects in gonadotropin production. J Clin Invest 1996;98:1055–1062.
30 Domenice S, Latronico AC, Brito VN, Arnhold IJ, Kok F, Mendonca BB: Adrenocorticotropin-dependent precocious puberty of testicular origin in a boy with X-linked adrenal hypoplasia congenita due to a novel mutation in the DAX1 gene. J Clin Endocrinol Metab 2001;86:4068–4071.
31 Ahmad I, Paterson WF, Lin L, Adlard P, Duncan P, Tolmie J, Achermann JC, Donaldson MD: A novel missense mutation in DAX-1 with an unusual presentation of X-linked adrenal hypoplasia congenita. Horm Res 2007;68:32–37.
32 Jeffs B, Meeks JJ, Ito M, Martinson FA, Matzuk MM, Jameson JL, Russell LD: Blockage of the rete testis and efferent ductules by ectopic Sertoli and Leydig cells causes infertility in Dax1-deficient male mice. Endocrinology 2001;142:4486–4495.
33 Mantovani G, De ME, Borretta G, Radetti G, Bondioni S, Spada A, Persani L, Beck-Peccoz P: DAX1 and X-linked adrenal hypoplasia congenita: clinical and molecular analysis in five patients. Eur J Endocrinol 2006;154:685–689.
34 Achermann JC, Ito M, Silverman BL, Habiby RL, Pang S, Rosler A, Jameson JL: Missense mutations cluster within the carboxyl-terminal region of DAX-1 and impair transcriptional repression. J Clin Endocrinol Metab 2001;86:3171–3175.
35 Lehmann SG, Wurtz JM, Renaud JP, Sassone-Corsi P, Lalli E: Structure-function analysis reveals the molecular determinants of the impaired biological function of DAX-1 mutants in AHC patients. Hum Mol Genet 2003;12:1063–1072.
36 Achermann JC, Silverman BL, Habiby RL, Jameson JL: Presymptomatic diagnosis of X-linked adrenal hypoplasia congenita by analysis of DAX1. J Pediatr 2000;137:878–881.
37 Tabarin A, Achermann JC, Recan D, Bex V, Bertagna X, Christin-Maitre S, Ito M, Jameson JL, Bouchard P: A novel mutation in DAX1 causes delayed-onset adrenal insufficiency and incomplete hypogonadotropic hypogonadism. J Clin Invest 2000;105: 321–328.
38 Mantovani G, Ozisik G, Achermann JC, Romoli R, Borretta G, Persani L, Spada A, Jameson JL, Beck-Peccoz P: Hypogonadotropic hypogonadism as a presenting feature of late-onset X-linked adrenal hypoplasia congenita. J Clin Endocrinol Metab 2002; 87:44–48.
39 Ozisik G, Mantovani G, Achermann JC, Persani L, Spada A, Weiss J, Beck-Peccoz P, Jameson JL: An alternate translation initiation site circumvents an amino-terminal DAX1 nonsense mutation leading to a mild form of X-linked adrenal hypoplasia congenita. J Clin Endocrinol Metab 2003;88:417–423.
40 Shaikh MG, Boyes L, Kingston H, Collins R, Besley GTN, Padmakumar B, Ismayl O, Hughes I, Hall CM, Hellerud C, Achermann JC, Clayton PE: Skewed X-inactivation is associated with phenotype in a female with adrenal hypoplasia congenita. J Med Genet 2008; in press.
41 Ito M, Yu R, Jameson JL: DAX-1 inhibits SF-1-mediated transactivation via a carboxy-terminal domain that is deleted in adrenal hypoplasia congenita. Mol Cell Biol 1997;17:1476–1483.
42 Lalli E, Sassone-Corsi P: DAX-1, an unusual orphan receptor at the crossroads of steroidogenic function and sexual differentiation. Mol Endocrinol 2003;17: 1445–1453.
43 Achermann JC, Ito M, Ito M, Hindmarsh PC, Jameson JL: A mutation in the gene encoding steroidogenic factor-1 causes XY sex reversal and adrenal failure in humans. Nat Genet 1999;22:125–126.
44 Achermann JC, Ozisik G, Ito M, Orun UA, Harmanci K, Gurakan B, Jameson JL: Gonadal determination and adrenal development are regulated by the orphan nuclear receptor steroidogenic factor-1, in a dose-dependent manner. J Clin Endocrinol Metab 2002;87:1829–1833.

45 Biason-Lauber A, Schoenle EJ: Apparently normal ovarian differentiation in a prepubertal girl with transcriptionally inactive steroidogenic factor 1 (NR5A1/SF-1) and adrenocortical insufficiency. Am J Hum Genet 2000;67:1563–1568.

46 Kim CJ, Lin L, Huang N, Quigley CA, AvRuskin TW, Achermann JC, Miller WL: Severe Combined Adrenal and Gonadal Deficiency Caused by Novel Mutations in the Cholesterol Side Chain Cleavage Enzyme, P450scc. J Clin Endocrinol Metab 2008;93: 696–702.

47 Baker BY, Lin L, Kim CJ, Raza J, Smith CP, Miller WL, Achermann JC: Nonclassic congenital lipoid adrenal hyperplasia: a new disorder of the steroidogenic acute regulatory protein with very late presentation and normal male genitalia. J Clin Endocrinol Metab 2006;91:4781–4785.

48 Vilain E, Le MM, Lecointre C, Desangles F, Kay MA, Maroteaux P, McCabe ER: IMAGe, a new clinical association of intrauterine growth retardation, metaphyseal dysplasia, adrenal hypoplasia congenita, and genital anomalies. J Clin Endocrinol Metab 1999; 84:4335–4340.

49 Bergada I, Del RG, Lapunzina P, Bergada C, Fellous M, Copelli S: Familial occurrence of the IMAGe association: additional clinical variants and a proposed mode of inheritance. J Clin Endocrinol Metab 2005;90:3186–3190.

50 Flück CE, Achermann JC, Miller WL: The adrenal cortex; in Sperling MA (ed): Pediatric Endocrinology, ed 3. Amsterdam, Elsevier, 2008, in press.

51 Rajab A, Kelberman D, de Castro SCP, Biebermann H, Shaikh H, Pearce K, Hall CM, Shaikh G, Gerrelli D, Grueters A, Krude H, Dattani MT: Novel mutations in LHX3 are associated with hypopituitarism and sensorineural hearing loss. Hum Mol Genet 2008; in press. DOI: 10.1093/hmg/ddn114

Note added in proof

Panhypopituitarism, including ACTH deficiency, has now been reported in individuals with mutations Lim homeobox gene 3 (LHX3) [51].

Dr. John C. Achermann
Developmental Endocrinology Research Group
Clinical and Molecular Genetics Unit, UCL Institute of Child Health, University College London
30 Guilford Street, London WC1N 1EH (UK)
Tel. +44 207 905 2887, Fax +44 207 404 6191, E-Mail j.achermann@ich.ucl.ac.uk

Flück CE, Miller WL (eds): Disorders of the Human Adrenal Cortex.
Endocr Dev. Basel, Karger, 2008, vol 13, pp 33–54

Adrenal Androgens in Humans and Nonhuman Primates: Production, Zonation and Regulation

Ann D. Nguyen · Alan J. Conley

Population Health and Reproduction, School of Veterinary Medicine,
University of California-Davis, Davis, Calif., USA

Abstract

The synthesis and secretion of large quantities of the adrenal androgens, dehydroepiandrosterone (DHEA) and its sulfoconjugate DHEA sulfate (DS), is a phenomenon that appears limited to humans and some nonhuman primates. Both hydroxylase and lyase activities of the enzyme 17α-hydroxylase/17,20-lyase cytochrome P450 (P450c17) are necessary for DHEA production and are differentially regulated during adrenal development. Production of DHEA and DS occurs in the zona reticularis (ZR) of adults and the fetal zone of fetal primate adrenal glands, which is the primary substrate for maternal estrogen production during pregnancy. The onset of adrenal androgen production in childhood, referred to as adrenarche, corresponds with the establishment of the ZR: but the process is poorly understood, largely due to the lack of accessible animal models. Several nonhuman primates have been used to study adrenal function and remodeling, though none completely recapitulates human adrenarche, developmentally, functionally or temporally. This review will summarize the variations in adrenal androgen production and adrenal zonation in humans and nonhuman primates throughout life. It is hoped that recent studies demonstrating adrenarche in the rhesus will put in proper context the significance of adrenal zonation in nonhuman primates as valid models for human adrenal development and function.

The mammalian adrenal cortex is remarkably zonal in its cellular organization and functional ability to secrete steroids. The major zones of the mature cortex, the outermost zona glomerulosa (ZG), and beneath it the zona fasciculata (ZF), secrete mineralocorticoids, and glucocorticoids, respectively. In addition, in humans and some nonhuman primates, the zona reticularis (ZR) located between the ZF and the medulla secretes androgens [1]. The secretion of adrenal androgens in primates is profoundly dynamic, reflecting the development of the ZR postnatally and also the functional capacity for androgen synthesis by the transient fetal zone (FZ) prenatally. In humans, dehydroepiandrosterone (DHEA) and its sulfoconjugate (DS) are the principal androgens secreted from the FZ and ZR, though these zones are distinctly different morphologically and in their appearance developmentally. This is reflected

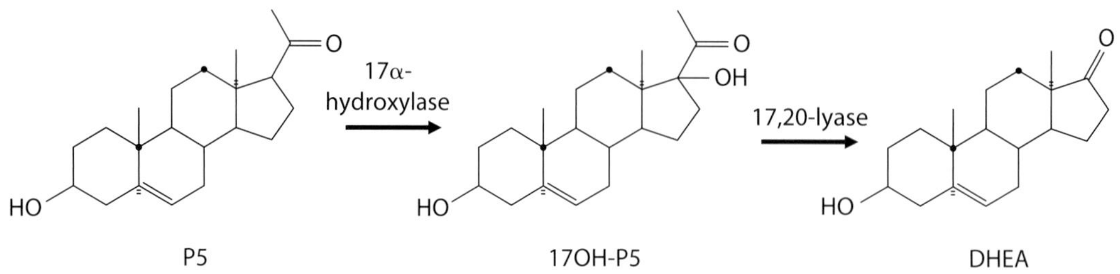

Fig. 1. Diagram showing metabolism of P5 by P450c17. Initial metabolism by 17α-hydroxylase activity of P450c17 produces 17OH-P5, which can be metabolized subsequently to DHEA by the 17,20-lyase activity of P450c17.

in circulating DHEA and DS levels that fluctuate greatly throughout pre- and postnatal development [2–6] with differentiation of the FZ during gestation [7, 8], or the ZR in adulthood [9, 10]. DHEA and DS levels decline with regression of the FZ after birth and then rebound around 5–7 years of age [2, 3] with the induction and establishment of the ZR [11]. This endocrinological event is referred to as adrenarche [12] and its onset is independent of the timing of puberty [13, 14]. Early studies characterizing changes in DS levels with age [15, 16] led to the generally held view that humans and chimpanzees are the only primate species which experience adrenarche. While the morphological and endocrine aspects of adrenal androgen output have been well characterized in humans [4, 11, 17], the cellular and molecular elements initiating and regulating ZR differentiation have yet to be clarified in any primate. Yet, an understanding of normal ZR development and factors inducing premature adrenarche may provide insight into the etiology of hyperandrogenism associated with diseases such as polycystic ovarian syndrome [18], or help in identifying antecedents of these conditions in early life. It is the authors' belief that nonhuman primates like the rhesus macaque experience the same remodeling of adrenal structure and function as humans and provide the only relevant animal models with which to study these phenomena.

Adrenal development and zonation can be defined in functional terms by the zonal expression of the enzymes that regulate steroid synthesis. The enzyme 17α-hydroxylase/17,20-lyase cytochrome P450 (P450c17) is central in this regard because it is responsible for the synthesis of DHEA from pregnenolone (P5), the precursor of all adrenal steroids, but is equally critical for cortisol synthesis. It is just as important that the 17deoxycorticoids, aldosterone and corticosterone, are synthesized in the absence of P450c17. If P5 is hydroxylated by P450c17 and the product, 17α-OH-P5, is oxidized in a second concerted reaction, C17–20 cleavage (17–20-lyase) follows and results in DHEA formation (fig. 1). 17α-OH-P5 (and/or 17α-OH-progesterone) is

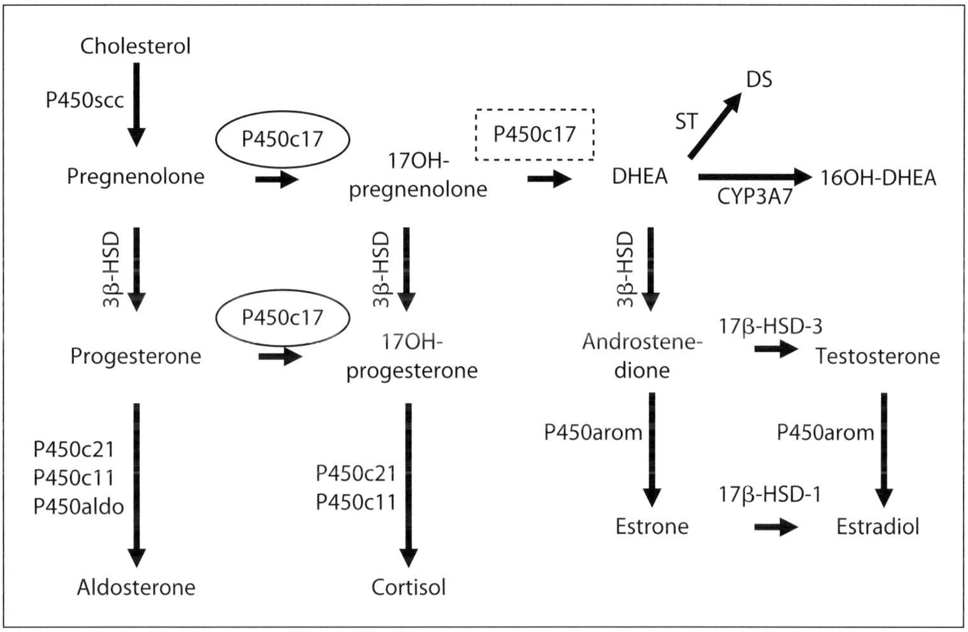

Fig. 2. Overview of the steroidogenic pathways resulting in mineralocorticoid (aldosterone), glucocorticoid (cortisol), and sex hormone synthesis in humans and nonhuman primates. Circled P450c17 represents 17α-hydroxylase activity of P450c17. Boxed P450c17 represents 17,20-lyase activity of the enzyme. Shown also are the enzymes: cholesterol side chain cleavage (P450scc), P450c21, P450c11, aldosynthase (P450aldo) and aromatase (P450arom) cytochromes P450, 3β-HSD, 17β-hydroxysteroid dehydrogenase types 1 (17β-HSD-1) and 2 (17β-HSD-2) and steroid sulphotransferase (ST).

also an essential substrate in the synthesis of cortisol (fig. 2), which is produced in the ZF wherein P450c17 catalyzes only the first oxygenation reaction. Even though P450c17 (like all P450s) has only a single substrate-binding pocket, the 17,20-lyase activity of the enzyme can be physiologically regulated independent of 17α-hydroxylase activity by several mechanisms: serine/threonine phosphorylation [19, 20], changes in availability of the essential redox partner binding flavoprotein cytochrome P450 oxidoreductase (CPR) [21], and/or the allosteric action of cytochrome b5 (b5) [22–24]. The role of b5 in the regulation of 17,20-lyase activity and androgen synthesis makes this enzyme a particularly useful differentiative marker of adrenal zonation and function. Moreover, the fundamental similarities in adrenal development that are shared among certain primate species suggest avenues for research in nonhuman primates to increase current understanding of the physiological regulation of adrenal androgen secretion. This review will summarize the present state of knowledge of adrenal zonal differentiation in humans, and the major nonhuman primate species studied to date, including the major developmental transitional periods and regulation of adrenal androgen synthesis.

Steroidogenic Pathways and Enzymes Involved in Adrenal Steroid Synthesis

The conversion of cholesterol to P5 sets the upper limit of the extent of steroid hormone production in tissues not receiving precursors from remote sources (e.g. maternal substrate supply to fetuses). However, metabolism of P5 by the microsomal enzymes P450c17 and/or 3β-hydroxysteroid dehydrogenase/$\Delta^{5,4}$-isomerase (3β-HSD), determines in large part the class of steroid that will be ultimately synthesized [25]. Expression of 3β-HSD in the adrenal cortex is necessary for the formation of progesterone (P4) which is metabolized by additional steroid-hydroxylating P450s including 21-hydroxylase (P450c21) and 11β-hydroxylase (P450c11) into the 17-deoxycorticoids corticosterone and aldosterone [26]. However, the production of both cortisol and sex steroids relies upon 17α-hydroxylation of P5 by P450c17 and subsequent metabolism by 3β-HSD. Through the activity of 3β-HSD, 17OH-P5, a Δ^5 substrate, can be converted to the Δ^4 product 17OH-P4. Further oxidation of 17OH-P4 by P450c21 and P450c11 results in the synthesis of cortisol, the major glucocorticoid secreted by the primate adrenal cortex. In the relative absence of 3β-HSD activity, 17OH-P5 is oxidized a second time and subsequently cleaved to the Δ^5 19 carbon 'androgen' DHEA (fig. 1) by the 17,20-lyase activity of P450c17 [17, 27, 28]. The 17,20-lyase activity of the human P450c17 enzyme represents a significant branch point in steroid hormone synthesis [25], because the cleavage of Δ^4 pregnanes is very inefficient by comparison to Δ^5 substrates [29]. In other words, the poor rate of cleavage of 17OH-P4 to androstenedione represents a relative metabolic block that prevents the efficient production of sex steroids from Δ^4 pregnanes. Therefore, P4 production by 3β-HSD effectively precludes metabolism through 17,20-lyase, and shunts the pathway away from DHEA and sex steroid synthesis (see fig. 2), which in the adrenal cortex effectively feeds the cortisol synthetic pathway.

Adrenal Androgen Production in Humans and Nonhuman Primates

Circulating concentrations of adrenal androgens vary with age [30] and across primate species. The majority of studies investigating human or nonhuman primate adrenal androgen production have employed cross-sectional sampling, examining serum levels of DHEA and/or DS [2, 3, 5, 6, 15, 16, 31–35]. These data form the foundation for much of our current understanding of adrenarche. Longitudinal studies on DS [36] and others examining urinary metabolites [37] highlight four dynamic developmental periods of adrenal androgen secretion in humans, the development and subsequent regression of the FZ, and the establishment (adrenarche) and senescence of the ZR. These developmental periods will be compared in humans and nonhuman primates in order to evaluate nonhuman primate species as models for human adrenal development.

Adrenal Androgen Production in Humans

In humans, umbilical cord plasma concentrations of DS are relatively constant throughout the late second and third trimesters, but experience a marked increase in late-gestation [5]. Measured concentrations of DS (about 2,400 ng/ml) are twofold greater near term compared to mid-gestation (about 1,200 ng/ml) [5]. DS levels rise markedly in the weeks immediately prepartum, and are correlated with a rapid increase in adrenal mass in the late gestation fetus [5]. Increasing levels of fetal DS are associated with diminishing cholesterol levels [38] and are mirrored in rising maternal estriol levels [5, 30]. Administration of exogenous DS to pregnant women increased circulating levels of estrone and estradiol [39], suggesting that maternal estrogen production reflects fetal adrenal androgen output. While the regulatory factors involved in stimulating adrenal androgen production in late gestation have not been identified, it is apparent that the fetal hypothalamic-pituitary-adrenal axis plays a major role. Specifically, anencephaly [40] and suppression of the fetal hypothalamic-pituitary axis [41] both greatly reduce fetal adrenal DS and maternal estrogen output. Both DHEA and cortisol secretion by the fetal adrenal are stimulated, though to different degrees, by ACTH [42]. This suggests that there is also differential regulation of secretion of these steroids within the adrenal cortex.

Postnatally, there is a dramatic decline in circulating DHEA and DS levels coincident with the involution of the human FZ [2, 3, 43] which occurs within the first months of life. Circulating DHEA and DS levels remain low through early childhood [2, 3, 44] until adrenarche. Adrenarche is defined clinically by a rise in circulating DS levels above 40–50 μg/dl [4], and is associated with the appearance of pubic and axial hair. Cross-sectional studies have found that DHEA and DS levels begin to rise between 5–7 years of age, with an earlier rise in girls than in boys [2, 3, 45]. However, recent longitudinal studies in both healthy children [46] and others suggest that adrenarche is not sudden, but is a gradual process and represents a steady increase in DS from as early as 3 years of age [36, 37]. This is consistent with results of cross-sectional studies on the urinary excretion of DHEA metabolites, such as androstenediol and 16OH-DHEA, which increased steadily with age during childhood and adolescence [47]. Efficient metabolism of DHEA in early childhood and adolescence may have confounded previous estimates of adrenal androgen output [47]. If so, changes in steroid metabolism must be specific for DHEA because there is no corresponding change in cortisol concentrations or other circulating steroids [4, 48]. It has been long postulated that specific adrenotropic factors distinct from ACTH are responsible for control of adrenal androgen secretion [49, 50]. Existence of cortical androgen-stimulating hormone and adrenal androgen-stimulating hormone has been long disputed. Particular regions of human proopiomelanocortin were thought to be cortical androgen-stimulating hormone 'candidates', though studies found no stimulation of steroidogenesis in adult and fetal adrenal cells exposed to human proopiomelanocortin-(79–96) [49, 51]. A 60-kDa peptide isolated from the human pituitary differentially stimulated secretion of DHEA by canine adrenals, with no apparent effect on cortisol production [50]. More definitive studies have not appeared and doubt persists as to the significance of these findings [52]. It is unlikely that ACTH plays a causative role in this phenomenon as both ACTH and cortisol levels remain steady during this developmental period [48]; however, ACTH is likely a permissive factor as it is necessary for normal adrenarche to occur [4]. Insulin-like growth factor-1 has been associated with increased androstenedione levels [53] in individuals that experienced premature adrenarche. How insulin-like growth factor-1 affects adrenal androgen production and what initiates adrenarche and ZR differentiation is unknown.

A gradual rise in adrenal androgens levels continues after adrenarche until peak levels are reached in the third decade of life (slightly earlier in females than males) [54]; adult DS levels of both sexes are greater than those observed in late gestation fetuses [30, 54]. After peak levels are reached in adulthood, there is a gradual decrease in circulating DS concentrations [6, 55]. This slow, lifelong decline in DS contrasts the dramatic postpartum decline which results from involution of the FZ. While an overall decline in DHEA and DS levels occurs with increasing age, a transient elevation in circulating DS is observed in some perimenopausal women from certain ethnic groups [56]. Once again, these changes in DHEA and DS secretion from the ZR are not mirrored in cortisol secretion [57], suggesting that factors other than ACTH are responsible [30]. While more specific changes in activities or expression of adrenal steroidogenic enzymes and accessory proteins may affect synthesis of DHEA and DS, reduction in the proportion of the adrenal cortex representing a functional ZR appears to be a factor in the age-related decline of androgen levels [58]. Immunohistochemical (IHC) studies examining reticular fibers and expression of b5 reported that there was a reduction in the width of the ZR, but no change in overall mass of the adrenal gland [59]. Further study of the factors which cause a reduction in DS levels are necessary for a better understanding of the regulation of adrenal androgen output.

Adrenal Androgen Production in Nonhuman Primates

Much of what is known about the development and function of fetal and adult adrenal development is inferred from circulating steroid levels, histological and immunocytochemical analysis, and is most clearly defined in the rhesus macaque. The majority of studies relating to developmental events rely on maternal estrogen levels as a surrogate for adrenal androgens; many fewer have examined fetal steroids or adrenal DHEA/DS output directly. Macaques have a functional fetoplacental unit for estrogen synthesis during pregnancy and levels of maternal estrogens (primarily estrone) [60] are dependent on androgen production by the fetal adrenal [61, 62]. Levels of maternal estrogens are lower in the rhesus compared to humans [1], likely limited by fetal DHEA and DS production, and perhaps confounded by conversion of DHEA to other metabolites [63]. The rhesus fetoplacental unit produces little estriol, due in part to low hepatic 16α-hydroxylase activity in this species [64]. The capacity for deconjugation seems not a limiting factor, as the rhesus placenta exhibits high

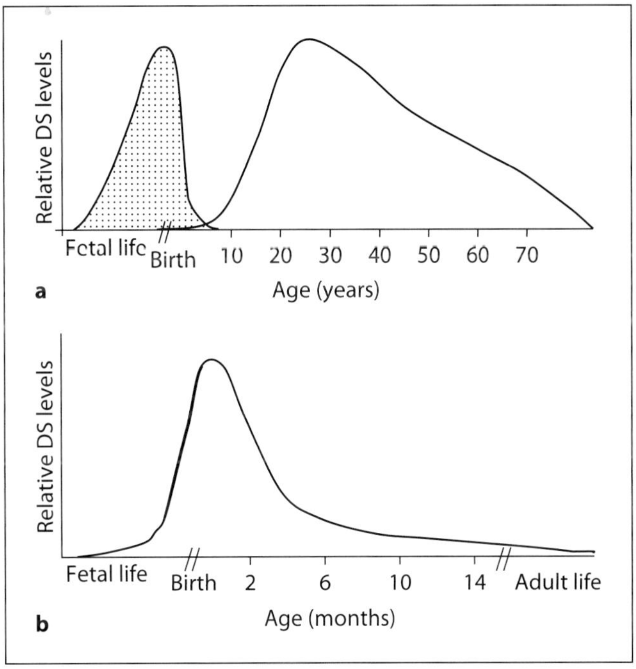

Fig. 3. Representative diagram of the DS profile in humans (**a**) and the rhesus (**b**) with major adrenocortical events demarcated. Blue areas indicate the chronological span of the active human FZ. Yellow areas indicate the chronological span of the active human ZR. Relative contributions by the rhesus FZ and ZR have not yet been determined.

sulfatase activity [65]. Circulating DS levels in the rhesus fetus remain low from early to mid-gestation and only rise in the immediate prepartum period [31, 66]. Changes in DS levels are mirrored in maternal estrone and estradiol levels, as well as fetal estrone levels [66]. This is likely driven in part by prepartum activation of the hypothalamic-pituitary axis. The hypothalamic-pituitary-adrenal axis is functional in the late-term rhesus, as dexamethasone administration suppresses fetal cortisol production [62] even though maternal treatment with ACTH fails to restore cortisol secretion in decapitated fetuses [67]. Both DS and cortisol are suppressed by dexamethasone in the second half of gestation [42]. Epidermal growth factor has also been shown to induce adrenocortical maturation in late-gestation fetal rhesus macaques [68].

The production of DS in neonates is probably comparable to that in late gestation, but there are much higher concentrations in neonates, in part due to a marked decrease in clearance rate after birth [69] that likely follows separation from the placenta. Few studies have carefully documented DS concentration in the early neonatal period of rhesus macaques. From the few neonatal samples examined as part of larger developmental studies, it appears that DS peaks in the first 2 months of life [31, 33]. Thereafter, adrenal androgen levels decline rapidly through the 1st year of life and continue to decrease throughout the rest of life [31–33]. Therefore, unlike the two temporally distinct peaks that characterize human adrenal androgen production (one in fetal life, and one between 20–30 years of age; fig. 3) the rhesus macaque exhibits only one peak, which occurs apparently in the perinatal period (fig. 3). The perinatal peak in DS has led some investigators to suggest that the rhesus FZ persists after birth [1, 69], in contrast to the human FZ which regresses postpartum. Moreover, administration of ACTH stimulates both DS and cortisol synthesis (presumably by the FZ and definitive zone – DZ, respectively) in the perinatal rhesus macaque, a response similar to that seen in the human fetus [70]. In this regard, adrenocortical function in the rhesus macaque might be seen to reflect that of the late-gestation human fetus. Adrenarche is not an event that has been recognized in the rhesus because there is no obvious increase of DHEA/DS secretion from 4 months to 3 years of age [15, 16] to mirror human adrenal maturation in childhood [1, 71].

The baboon has been used to study endocrinology of pregnancy and the factors which can regulate or can alter maternal estrogen secretion [72]. As with humans and several other nonhuman primates, maternal estradiol levels increase throughout gestation [73]. The baboon fetal adrenal supplies the precursors for placental production of estradiol and estrone. Like the rhesus macaque, estriol is not detectable in maternal plasma [74]. Fetal production of DHEA is stimulated by ACTH (and prolactin) during mid- and late gestation [35, 75, 76]. The response to ACTH is inhibited by estradiol, though estradiol alone does not have any apparent effect on DHEA secretion by the fetal adrenal cells [76]. However, estrogen has been shown to inhibit FZ expansion based on reduced FZ growth with pregnant baboons treated with letrozole, a potent and specific inhibitor of aromatase [77]. It is postulated that these effects are mediated by estrogen receptor-α (ERα) and estrogen receptor-β (ERβ), both of which are expressed throughout the fetal adrenal cortex, though most prominently in the DZ [78]. Fetal adrenocortical cells produce 10 times more DHEA than cortisol near term and production of DHEA and DS continues after birth for a few days, followed by a precipitous decline by the 2nd week of life [75]. During this time, the innermost FZ diminishes in width though the overall mass of the adrenal gland does not appreciably change [75]. Although these events parallel those of human fetal adrenal development, like the rhesus (and cynomolgus) macaque, the baboon is not believed to experience an adrenarche after birth.

Marmosets are another nonhuman primate species that has been used extensively in studies of stress, social structure and subordination, as well as reproductive physiology [79] whose fetal adrenal development and differentiation may well resemble those of human fetuses [80]. Once again, much of what is understood is inferred from circulating maternal estrogens or urinary estrogen excretion during pregnancy [1]. Estriol levels are low throughout pregnancy, though levels do increase through late gestation [81]. Maternal plasma estradiol and estrone levels increase through mid-gestation; in contrast to humans, estradiol is the major estrogen product of the marmoset placenta [81, 82]. Interestingly, urinary secretion of estradiol in the marmoset far exceeds secretion of estriol by humans [1]. Given the necessity of fetal DHEA and DS for placental production of estrogens, these studies suggest that the marmoset FZ is very active, at least during the last two thirds of gestation [80]. While direct studies of the fetal marmoset adrenal are lacking, the neonate has been used in studies examining zonation and steroidogenesis. Plasma DHEA and DS concentrations are elevated in the neonatal marmoset, over 20-fold greater than levels observed in adulthood [34]. Similar to humans and baboons, adrenal androgen levels rapidly decline after birth coincident with regression of the FZ [34, 80]. In fact, DHEA levels drop by over 95% in both males and females in the 1st year of life [34]. Although male marmosets and dominant females do not express a functional ZR in adulthood [80, 83], studies in marmosets may be useful in better understanding human fetal adrenal development and regression after birth. The apparent regeneration of some functional ZR tissue following ovariectomy [83] is particularly notable because there are few, if any, reports relating to experimental manipulations that can induce regeneration of ZR in mature primates.

It is a commonly held belief that the chimpanzee is the only nonhuman primate that exhibits an adrenarche comparable to that of humans [15, 71]. Chimpanzees do experience a rise in DHEA and DS levels at about 2 years of age before any increase in gonadal steroids can be detected in association with puberty [16], which occurs at about 5–6 years of age [15, 16]. However, these two cross-sectional studies are the only ones conducted thus far to investigate adrenarche in this species and no histological or biochemical evidence supportive of adrenarche has been obtained in chimpanzees. Maturation of a ZR, the hallmark of adrenarche [12], has yet to be examined in this Old World primate [1]. Longitudinal endocrine studies coupled with biochemical and histological investigations are necessary to clearly define adrenarche in the chimpanzee, as in other nonhuman primates in which adrenarche has not yet been observed.

The Primate Adrenal Cortex

Initial observations on the human adrenal date as far back as the 18th century, when Morgagni noted the relatively small mass of suprarenal glands of anencephalic fetuses [reviewed in 84], but details on its function remained unclear well into the 20th century. The involution of the FZ that was initiated at birth [43, 85] fueled speculation on its necessity for proper development in utero [86]. The pioneering studies of W.C. Hill were among the first to investigate the adrenal morphology of various classes of mammals, including nonhuman primates. He suggested that the rhesus macaque experienced involution of an FZ of the adrenal gland similar to humans, perhaps providing a useful model for adrenal morphology and development [84]. The androgenic function of the adult adrenal cortex was postulated by Grollman [87], who defined an androgenic zone of cells distinct from the remainder of the adrenal cortex. Early observations of adrenal hyperplasia, which was associated with virilization of females, as well as lack of gametogenesis in afflicted males, were taken as evidence in support of this conclusion [87]. The eventual quantification of androgens in adrenal venous blood in adult humans [9] further supported the existence of an androgenic zone distinct from the cortical zones responsible for corticoid synthesis.

Zonation of the Human Adrenal

The adult adrenal cortex synthesizes and secretes aldosterone, cortisol, and DHEA from the ZG, ZF and ZR, respectively. In contrast, the fetal adrenal cortex consists mostly of an FZ that primarily produces DHEA and DS, with little cortisol or aldosterone synthesis from the outermost DZ. The differences in hormonal output of the human adult and fetal adrenal cortex are largely a reflection of the steroidogenic enzyme profiles that determine the steroid products of each zone and define the distinctive morphological regions through development.

Attention was initially drawn to the human fetal adrenal by the rapid involution of the 'nonpermanent' fetal cortex (now termed the FZ) after birth and the remarkable size of the gland, proportionately 20 times larger than the adult adrenal gland [86]. A 50% reduction in the mass of the FZ (composed of large, fatty cells) was estimated to occur in the first few weeks of life [43]. While the earliest studies of FZ involution, conducted by Ewis and Pappenheimer, suggested that collapse of the FZ involved massive necrosis and hemorrhage [86], later studies did not corroborate these findings [43, 85, 88]. The FZ regresses gradually, concomitant with proliferation of the DZ and cells in a 'transitional' zone (TZ) beneath it, differentiating into the ZG and ZF, respectively, of the adult cortex [43]. The precise mechanism by which the FZ regresses is still unknown. Regression occurs in anencephalic fetuses [43, 87], and more rapidly in the absence of pituitary support, presumably ACTH [89].

Cell culture and advances in immunohistochemistry (IHC) have clarified both zone-specific secretory products and steroidogenic enzyme phenotypes of the cells comprising the human fetal and adult adrenal cortex. Metabolism of P5 by cultured human FZ cells results in the production of predominantly Δ^5-sulfate steroids, such as DS [7, 90–92]. Steroidogenic enzyme expression determined by IHC and in situ hybridization [93, 94] is consistent with an FZ that is capable of adrenal androgen production throughout the latter half of gestation. In cultured cells from the FZ of mid-gestation fetuses, DHEA and DS comprised roughly 90% of the total product of metabolism of P5, while cortisol accounted for only 8% [92] and aldosterone synthesis was undetectable [8, 90, 92]. The FZ expresses the steroidogenic enzymes necessary for DHEA synthesis including P450c17 and the accessory proteins, b5 and CPR, from 14 weeks of gestation onward [93–95]. In addition, cholesterol side chain cleavage (P450scc), DHEA sulfotransferase (DST), steroidogenic acute regulatory protein (StAR), and 21-hydroxylase cytochrome P450 (P450c21; see fig. 2 for pathway placement) were all immunodetectable during this developmental period [93, 94]. In all, these data are consistent with an FZ capable of synthesizing DHEA and DS from cholesterol early in gestation, a capacity that is retained to term. Production of DHEA and DS by the FZ has also been attributed to a diminished level of 3β-HSD expression and activity [7, 8, 93, 94] within this region. 3β-HSD expression remains essentially

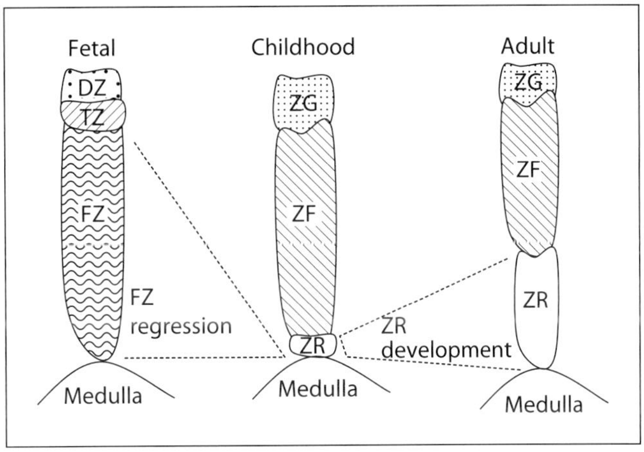

Fig. 4. Diagrammatic representation of human adrenocortical zonation at various stages of development. The fetal adrenal cortex is composed of three recognized regions: the DZ, TZ, and FZ. The FZ regresses after birth, and adult zonation, comprising the ZG, ZF and ZR becomes established. In childhood, the cortex is comprised mostly of the ZF, with a distinct ZG and limited ZR. The mature adrenal cortex exhibits a prominent ZR by comparison.

undetectable in the FZ, either by IHC or Western immunoblot analysis, throughout gestation [8, 91, 94, 96], though low levels of activity have been reported [7]. Measured 3β-HSD activity in the FZ cells is lower than in the neocortex [7], and expression throughout the fetal adrenal itself is <1% of adult levels [97]. Diminished competition between P450c17 and 3β-HSD for substrate, either due to reduced activity or expression, is postulated to further facilitate the production of DHEA and DS from the FZ [25].

Morphological assessments of the DZ, formerly referred to as the 'permanent zone' or neocortex [39], and its steroidogenic output, suggested that the adult cortex differentiates entirely from this region [8, 88]. In fact, early studies defined the FZ based on its location between the 'adult cortex' and the medulla [84]. However, more careful examination revealed the existence of a third zone both in the human (fig. 4) and the rhesus fetal adrenal (fig. 5, 6) [93]. Based on the observed similarity of steroidogenic enzyme profiles of mid-gestation and adult adrenals, the TZ has been hypothesized to differentiate into the ZF and the DZ into the ZG [93]. As with the FZ, the function of the neocortex was assessed in part based on investigations of fetal adrenal cell culture; however, earlier studies were unable to distinguish TZ and DZ cells as distinct adrenocortical regions [7, 90, 92]. The primary product of both P4 and P5 metabolism by the neocortex is cortisol, with some synthesis of corticosterone and deoxycortisol also detectable [7, 8, 90, 92]. Aldosterone production is not believed to occur normally in the mid-gestation fetal adrenal and is generally undetectable [92], though some can be found after incubating tissues with excess corticosterone [7]. Expression of 3β-HSD is induced in the DZ and TZ at mid-gestation and expression persists throughout late gestation. Although P450c17 and DST are expressed in the TZ throughout gestation, both are conspicuously absent from the DZ during all developmental periods [94]. StAR, P450scc, CPR, and b5 were all detected in the TZ throughout pregnancy,

Fig. 5. Diagrammatic representation of transformation of the fetal into the mature adrenal cortex of rhesus macaques. Like in humans, the fetal adrenal cortex is composed of the DZ, TZ and FZ. In addition, a 'dense band' of cells exists beneath the TZ that appears to give rise to the ZR in rhesus. The adrenal cortex of adult rhesus macaques possesses a large distinct ZR, in addition to prominent ZF and smaller ZG.

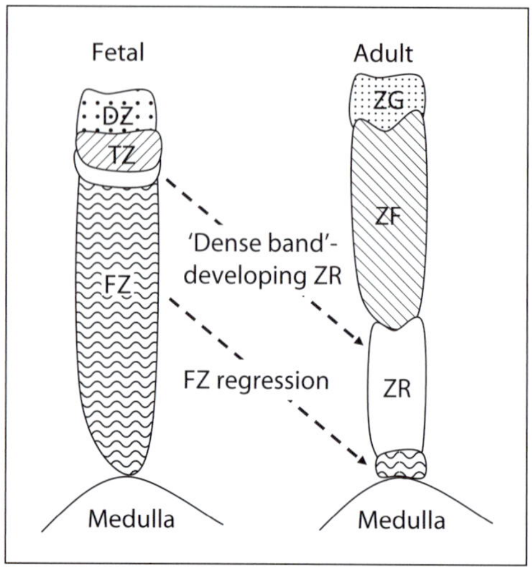

Fig. 6. Zonal differentiation of the mature rhesus adrenal cortex as shown by immunohistochemistry for key steroidogenic enzymes. The expression of b5, 3β-HSD, P450c17 and NADPH CPR is shown in the ZG, ZF and ZR. Bar represents 50 μm. Panels are the same region and same magnification.

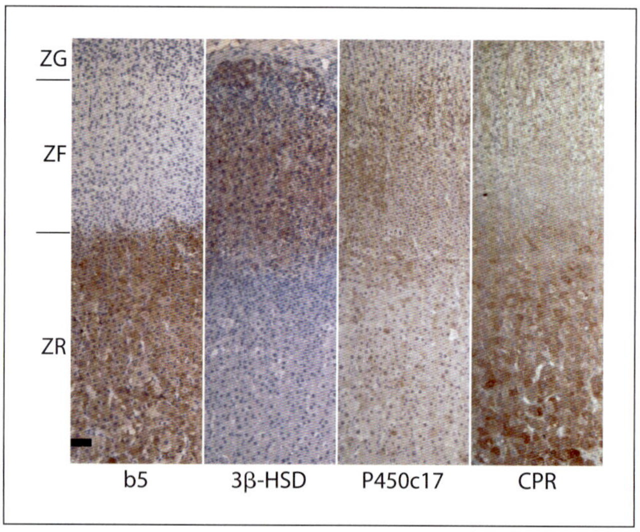

though not in the DZ until after 23 weeks of gestation [93, 94]. P450c21 was detected in the TZ and FZ as early as 13 weeks of gestation [98]. These data suggest that the TZ is responsible for the production of cortisol, as it expresses the steroidogenic enzymes necessary for glucocorticoid production, namely P450c21, P450c17, and 3β-HSD (from mid-gestation onward) [8, 93, 94, 98, 99]. Although aldosterone synthase cytochrome P450 is expressed in the fetal adrenal at early gestation, expression is

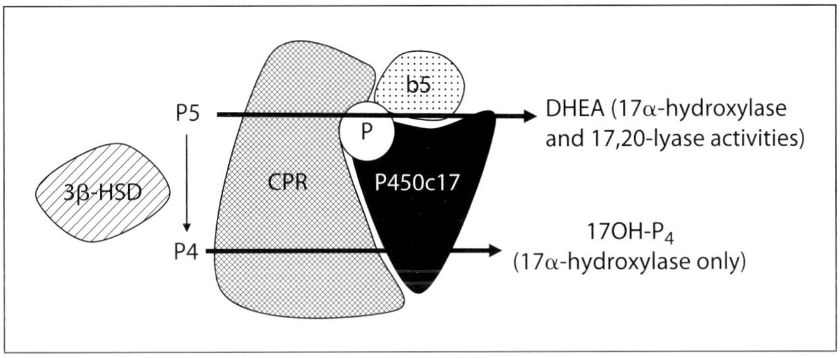

Fig. 7. Diagrammatic representation of the three major mechanisms believed to modulate the 17,20-lyase activity of the enzyme P450c17: the action of the accessory protein b5, relative levels of the redox partner CPR, and serine phosphorylation (circled P) of P450c17. 17,20-lyase activity is further promoted by the relative absence of competition and metabolism of P5 to P4 by 3β-HSD, because 17OH-P4 is not cleaved efficiently to the C19 (androgen) product, androstenedione.

confined to the FZ and TZ [98], both of which express P450c17 and lack 3β-HSD expression during that period. Additionally, StAR, P450scc, P450c21, CPR and 3β-HSD are all expressed in the DZ during which time there is little, if any, expression of P450c17 [94].

Postnatally, the human fetal adrenal cortex undergoes extensive remodeling, with differentiation of the TZ and DZ and regression of the FZ (fig. 4). As stated previously, the best morphological descriptions of FZ regression suggest that it is a gradual event culminating in an estimated 50% reduction in the mass of the FZ within the first few weeks of life, leaving a small band of connective tissue in its place [43, 85]. Studies of zonation in infant, juvenile, and adult adrenal glands are few and limited to specimens collected at postmortem, notably from deaths resulting from sudden traumas. The IHC studies of Suzuki et al. [11] are the most complete overview of steroidogenic enzyme expression at all stages of adrenal development, including perhaps the only comprehensive examination of tissues from infants. The once prominent FZ is reduced to a region that is only a few cell layers wide during infancy [11]. This is remarkable, given that the FZ accounts for as much as 80% of glandular mass during gestation [86]. 3β-HSD is localized throughout the ZF and ZG at 2 months old, but is absent from the ZR [96]. Decreased expression of P450c17, DST, b5, and CPR throughout the adrenal cortex [11] is coincident with the reduction in human FZ mass. Expression of P450c17, b5, and DST in the ZR is barely detectable from the ages of 7 months old to 5 years old, though 3β-HSD expression is high in all regions of the cortex during this developmental period [11, 96]. In fact, the ZR of infancy and early childhood possesses a steroidogenic enzyme profile similar to that of the ZF. Differentiation of a functional ZR is reported to be first visible at 5 years old with an increase in P450c17, b5, and CPR expression, and by 8 years old, there is a continuous region of reduced 3β-HSD expression located adjacent to the medulla [8, 10, 11, 17], another hallmark of a functional ZR. This morphological transformation occurs coincident with the rise in DHEA and DS levels that is used to define adrenarche in purely endocrinological terms. The ZG and ZF both express 3β-HSD along with CPR, which is necessary for the function of both P450c17 and P450c21. The ZF expresses P450c17, consistent with cortisol production [11]. Cytochrome b5 is absent from both the ZG and ZF of the adult adrenal cortex [11, 59, 95]. Studies specifically characterizing steroidogenic enzyme expression suggest it is more likely that the cortex reaches maturity around 8 years of age, though some have suggested the definitive adult cortex is not established until 11 years old [88] based on routine histology.

The factors which initiate the induction of the ZR have not yet been identified, nor are those responsible for the age-related decrease in ZR volume. As noted previously, DHEA and DS levels generally decline following a peak in the second or third decade of life [6, 54]. Some women also experience a transient increase in DS levels during the menopause transition, though not quite reaching levels seen during their peak at 20–30 years of age [56]. The gradual decline may be attributed to a progressive decrease in the number of functional ZR cells [57] or in ZR mass based on morphology and assessment and the expression of b5 [58, 59], an effective marker of ZR differentiation and androgenic capacity [100]. Cellular levels of b5 itself did not diminish with age, but the volume of the b5-positive region decreased by comparison to the ZG and ZF [59]. Further study is necessary to elucidate the mechanism by which this decrease in ZR mass occurs and to determine if the factors are involved in FZ regression. Similarly, detailed investigation of the perimenopausal rise in circulating levels of DS [56] is necessary to determine the link between the ovary and adrenal, factors which stimulate this 'reawakening', and if these are the same factors responsible for the onset of adrenarche.

The Nonhuman Primate Adrenal Cortex

Nonhuman primates experience very different patterns of adrenal development, and generally much shorter overall life spans. Therefore, it is unlikely that any nonhuman primate species experiences all stages, and the same temporal progression, of adrenal development and differentiation as humans. Some exhibit a morphologically prominent FZ during fetal development, but not with a comparable level of androgen secretion. Similarly, those known to possess a morphological ZR have circulating DS concentrations that are one tenth or less of human levels and adrenarche has not been documented by observation of an increase in adrenal androgen output. Differences in levels notwithstanding, the ZR of the human and at least some nonhuman primates is androgen secreting and the FZ regresses after birth, despite the different temporal patterns. This suggests that differences are more quantitative than qualitative, and that studies in nonhuman primates are instructive of human adrenal development and function.

The neonatal adrenal cortex of the common marmoset *(J. jaccus)* appears to share endocrinological similarities to the mid- and late-gestation human adrenal which is reflected equally in its steroidogenic enzyme profile. At 1 day old, the male marmoset fetal adrenal cortex exhibits three distinct regions, with a well-defined innermost FZ which exhibits marked expression of b5, P450c17, and CPR and reduced 3β-HSD levels, compared to adults [80]. These studies corroborate the earlier findings of Levine et al. [34], which were based on hematoxylin and eosin staining alone. 3β-HSD is expressed throughout the ZF and ZG of the neonatal adrenals (1 day old) as well as the entirety of the mature adrenal cortex in males [80]. P450scc, P450c21, and CPR are also expressed throughout all regions of the adult male adrenal cortex and, consistent with studies in other species, P450c17 was not detected in the ZG. In contrast to humans (and the rhesus), b5 expression was not detectable in any region of the mature marmoset adrenal cortex [80]. Early morphological studies noted the difficulty in identifying the ZR in adult marmosets [101]. The lack of a functional ZR corresponds with the low basal levels of circulating DHEA and DS [34], which fail to respond to ACTH stimulation [80]. However in female marmosets, recent studies suggest that social and gonadal status have a marked effect on adrenal zonation. The adult female subordinate marmoset has been shown to express a functional ZR and the ZR becomes more prominent after ovariectomy, based on b5 expression [83]. DHEA secretion in response to stimulation by ACTH is also augmented in ovariectomized females [83]. In this regard, the marmoset may be valuable in the study of adult adrenal development and induction of the functional ZR, which in humans effectively constitutes adrenarche.

Many studies have investigated and established the presence of a functional feto-placental unit in pregnant baboons. The baboon fetal adrenal produces higher levels of DHEA than cortisol and undergoes FZ regression following birth [75]. In accordance with this androgenic function, the baboon FZ expressed P450c17, as does the TZ which developed during late gestation [102]. Expression of 3β-HSD was confined to the DZ and TZ and undetectable throughout the FZ during mid- and late gestation [102]. These findings suggest that the FZ and TZ alone are capable of androgen and glucocorticoid production. More detailed study of the steroidogenic enzymes required for sex steroid, glucocorticoid, and mineralocorticoid production are necessary to properly characterize adrenocortical zonation of the fetal baboon. However, as noted above, inhibition of estrogen synthesis by administration of an aromatase inhibitor to pregnant baboons inhibited growth of the FZ, which was restored by estrogen supplementation [77]. In addition, ERα and ERβ, are expressed throughout the fetal adrenal cortex [78]. The expression of ERα and ERβ, though not exclusive to or even higher in the FZ, still supports a direct role of estrogen on FZ development and function [102]. These data are unique in offering possible mechanism of regulation of fetal adrenal output or responsiveness to ACTH and are among the very few studies to have done so. Much less is known about the establishment of a mature ZR in this species.

Like the baboon, studies of adrenocortical development and function in the rhesus macaque have focused primarily on fetal zonation. These include the landmark contributions of McNulty and colleagues which were significantly extended by Coulter, Mesiano and Jaffe. Rhesus fetal adrenal output of DHEA and DS is low until near term [31] and the majority of studies conducted with the fetal rhesus have examined this late gestational period. Some studies examining early gestation found faint expression of b5 and P450c17 in the FZ; however, expression of both proteins diminished and remained low until about 150 days of gestation [100]. A marked increase in b5 expression was observed in the FZ, along with increased P450c17 in the TZ [100], which likely coincides with rising DHEA and DS levels in late gestation. The rhesus fetal adrenal expresses P450c17 throughout the FZ and TZ during late gestation, indicating that the late gestational cortex is capable of cortisol and adrenal androgen production under normal, physiological conditions [93, 100]. Additionally, 3β-HSD expression was not detectable in the FZ at any stage of gestation, though it is present in both the TZ and DZ [103]. P450scc was expressed throughout the adrenal cortex, as was P450c21 [93, 98]. The importance of P450c21 expression in the FZ is unknown [98], but in the relative absence of 3β-HSD activity, it seems unlikely that it could significantly impact androgen synthesis. It is noteworthy, however, that the expression of enzymes required for androgen production, P450c17 and b5, was particularly high in a dense band of cells underlying the TZ immediately prepartum [100]. This region was characterized morphologically by McNulty et al. [104], who noted that the rhesus FZ appeared to gradually regress underneath this widening 'dense band'. Furthermore, they suggested that the 'dense band' of cells underlying the developing ZF differentiates into the mature ZR, and this occurs coincident with FZ regression (fig. 5) [104]. Early morphological studies suggested that the rhesus macaque experienced involution of the FZ similar to humans [84]. Others have postulated that the adult ZR, ZF, and ZG develop from corresponding regions of the fetal adrenal cortex (FZ, TZ, and DZ) based on steroidogenic enzyme profiles and production of DHEA and cortisol [93], but the origins of the zones of the mature adrenal cortex remain unclear.

To date, McNulty and colleagues are the only investigators to have examined the structure and zonation of the rhesus adrenal during the early neonatal period. Early morphological studies attributed the width of the rhesus adult adrenal cortex to a thick band of cells which corresponded to the human ZR [84]. Subsequent investigations confirmed that the rhesus macaque is one of a few primates known to exhibit a functional ZR in adult life [1]. The rhesus ZR exhibits a zonation and steroidogenic enzyme distribution (fig. 6) similar to humans, consisting of an innermost ZR with high expression of DST [105], b5 and P450c17, while lacking 3β-HSD expression [106]. P450c17 is also expressed in the ZF, consistent with its role in cortisol production. 3β-HSD is expressed at high levels in the ZF and ZG [106]. The redox partner CPR was localized throughout the adrenal cortex, with the greatest intensity of expression near the corticomedullary border [106]. DST expression was also most pronounced in the reticularis area nearest the medulla, though expression was uneven within the defined ZR [105]. Therefore, it is likely that the region nearest the medulla represents the area of greatest DS output, given the high expression of both DST and CPR. Further studies are necessary to determine the steroid production of this region and if the effects of aging in the human ZR are paralleled in the rhesus ZR.

The late-gestation rhesus has been used in the past as a model for human adrenal differentiation in late gestation and all evidence thus far suggests that this is justified on both physiological and morphological grounds. However, doubts have been expressed that rhesus macaques experience an adrenarche postnatally. Previous studies measuring DS levels in systemic plasma have failed to demonstrate a pre-pubertal rise like that occurring during human development (by definition, 'adrenarche') [15, 16]. Still, the rhesus clearly possesses both a morphologically distinct and functionally competent FZ in late gestation [93, 100] as well as a mature ZR in adulthood [106]. In this way, rhesus adrenocortical differentiation must be similar to that in humans, even if it occurs over a much shorter time frame and essentially during perinatal development. Recent studies completed in our laboratory provide a detailed analysis of adrenocortical differentiation in neonatal rhesus macaques ranging in age from 1 day to 1.3 years old. Cortical maturation was examined morphologically using IHC to define expression of key steroidogenic enzyme and accessory proteins, including b5, and was confirmed biochemically by immunoblot and enzyme activity assays. Our IHC data support the proposal by McNulty et al. [104] that a population of cells forming a 'dense band' beneath the TZ (but distinct from the cells of the FZ) differentiates into the mature ZR (fig. 5). The rhesus FZ and ZR exhibit different steroidogenic phenotypes and can be distinguished based on distinct cellular morphologies, aided by expression of P450c21. FZ cells are typically large and vacuolated, and lack P450c21 expression during the perinatal period. ZR cells are compact, densely packed and clearly express P450c21. The results of these analyses indicate that the rhesus FZ regresses during the prenatal and neonatal periods, an event that is entirely coincident with ZR differentiation. The coincidence of these two events in the rhesus is important mechanistically. It suggests that either the factors responsible for inducing FZ regression have the opposite effect, promoting ZR differentiation, or entirely different factors are recruited. Second, it emphasizes the distinct origins of these two cell types. In this, the results also highlight the value of studying nonhuman primate species.

The biochemical differentiation of the rhesus ZR (functional adrenarche) was also investigated between birth and 3 months of age by measuring the capacity for androgen synthesis (17,20-lyase) and assessing the expression of P450c17, b5, 3β-HSD and P450c21 in adrenal microsomes by immunoblot analysis. There was a linear rise in rhesus adrenal 17,20-lyase activity from birth that peaked between 2 and 3 months old. This was positively correlated with an increase in expression of b5, but there was no change in P450c17 or P450c21 expression. In addition, the importance of changes in b5 levels as a major determinant in 17,20-lyase activity was investigated by supplementing adrenal microsomes with recombinant b5. Addition of b5 stimulated

17,20-lyase activity (and DHEA synthesis) in adrenal microsomes with the lowest endogenous expression levels, restoring activity to levels seen in samples with high b5 expression. These data suggest that b5 is a key element in limiting androgen synthesis by the developing ZR and are the first to provide insight into the regulation of 17,20-lyase activity at physiological levels of enzyme expression. Moreover, the increase in 17,20-lyase activity was associated with a slight increase in 3β-HSD expression. Collectively, the IHC and biochemical studies indicate that the rhesus experiences adrenarche, that b5 is a driving factor in the phenomenon, and that a decrease in 3β-HSD expression is not obligatory. Furthermore, they support the idea that adrenarche is a gradual process, though completed during a compressed interval of early neonatal life in rhesus macaques, coincident with FZ regression. Lastly, these observations suggest that, because the regression of the FZ and establishment of the ZR are coincident events in the rhesus macaque, it is not possible to distinguish these events by systemic profiles of adrenal androgens. In other words, the neonatal increase in DS levels in rhesus macaques is not due to a persistent FZ, but to the establishment of a mature ZR that represents adrenarche in this species.

References

1 Conley AJ, Pattison JC, Bird IM: Variations in adrenal androgen production among (nonhuman) primates. Semin Reprod Med 2004;22:311–326.

2 de Peretti E, Forest MG: Unconjugated dehydroepiandrosterone plasma levels in normal subjects from birth to adolescence in human: the use of a sensitive radioimmunoassay. J Clin Endocrinol Metab 1976;43:982–991.

3 de Peretti E, Forest MG: Pattern of plasma dehydroepiandrosterone sulfate levels in humans from birth to adulthood: evidence for testicular production. J Clin Endocrinol Metab 1978;47:572–577.

4 Auchus RJ, Rainey WE: Adrenarche – physiology, biochemistry and human disease. Clin Endocrinol 2004;60:288–296.

5 Parker CR Jr, Leveno K, Carr BR, Hauth J, MacDonald PC: Umbilical cord plasma levels of dehydroepiandrosterone sulfate during human gestation. J Clin Endocrinol Metab 1982;54:1216–1220.

6 Orentreich N, Brind JL, Vogelman JH, Andres R, Baldwin H: Long-term longitudinal measurements of plasma dehydroepiandrosterone sulfate in normal men. J Clin Endocrinol Metab 1992;75:1002–1004.

7 Simonian MH, Capp MW: Characterization of steroidogenesis in cell cultures of the human fetal adrenal cortex: comparison of definitive zone and fetal zone cells. J Clin Endocrinol Metab 1984;59:643–651.

8 Doody KM, Carr BR, Rainey WE, Byrd W, Murry BA, Strickler RC, Thomas JL, Mason JI: 3ß-hydroxysteroid dehydrogenase/isomerase in the fetal zone and neocortex of the human fetal adrenal gland. Endocrinology 1990;126:2487–2492.

9 Wieland RG, Decourcy C, Levy RP, Zala AP, Hirschmann H: C-19-O-2 Steroids and Some of Their Precursors in Blood from Normal Human Adrenals. J Clin Invest 1965;44:159–168.

10 Endoh A, Kristiansen SB, Casson PR, Buster JE, Hornsby PJ: The zona reticularis is the site of biosynthesis of dehydroepiandrosterone and dehydroepiandrosterone sulfate in the adult human adrenal cortex resulting from its low expression of 3ß-hydroxysteroid dehydrogenase. J Clin Endocrinol Metab 1996;81:3558–3565.

11 Suzuki T, Sasano H, Takayama J, Kaneko C, Freiji WA, Carr BR, Rainey WE: Developmental changes in steroidogenic enzymes in human postnatal adrenal cortex: immunohistochemical studies. Clin Endocrinol 2000;53:739–747.

12 Havelock JC, Auchus RJ, Rainey WE: The rise in adrenal androgen biosynthesis: adrenarche. Semin Reprod Med 2004;22:337–347.

13 Sklar CA, Kaplan SL, Grumbach MM: Evidence for dissociation between adrenarche and gonadarche: studies in patients with idiopathic precocious puberty, gonadal dysgenesis, isolated gonadotropin deficiency, and constitutionally delayed growth and adolescence. J Clin Endocrinol Metab 1980;51:548–556.

14 Counts DR, Pescovitz OH, Barnes KM, Hench KD, Chrousos GP, Sherins RJ, Comite F, Loriaux DL, Cutler GB Jr: Dissociation of adrenarche and gonadarche in precocious puberty and in isolated hypogonadotropic hypogonadism. J Clin Endocrinol Metab 1987;64:1174–1178.
15 Smail PJ, Faiman C, Hobson WC, Fuller GB, Winter JS: Further studies on adrenarche in nonhuman primates. Endocrinology 1982;111:844–848.
16 Cutler GB Jr, Glenn M, Bush M, Hogden GH, Graham CE, Loriaux DL: Adrenarche: a survey of rodents, domestic animals, and primates. Endocrinology 1978;103:2112–2118.
17 Gell JS, Carr BR, Sasano H, Atkins B, Margraf L, Mason JI, Rainey WE: Adrenarche results from development of a 3ß-hydroxysteroid dehydrogenase-deficient adrenal reticularis. J Clin Endocrinol Metab 1998;83:3695–3701.
18 Miller WL: The molecular basis of premature adrenarche: an hypothesis. Acta Paediatr Suppl 1999; 88:60–66.
19 Zhang LH, Rodrigues H, Ohno H, Miller WL: Serine phosphorylation of human P450c17 increases 17,20-lyase activity: implications for adrenarche and the polycystic ovary syndrome. Proc Natl Acad Sci USA 1995;92:10619–10623.
20 Pandey AV, Miller WL: Regulation of 17,20 lyase activity by cytochrome b5 and by serine phosphorylation of P450c17. J Biol Chem 2005;280:13265–13271.
21 Yanagibashi K, Hall PF: Role of electron transport in the regulation of the lyase activity of C21 side-chain cleavage P450 from porcine adrenal and testicular microsomes. J Biol Chem 1986;261:8429–8433.
22 Katagiri M, Kagawa N, Waterman MR: The role of cytochrome b5 in the biosynthesis of androgens by human P450c17. Arch Biochem Biophys 1995;317: 343–347.
23 Onoda M, Hall PF: Cytochrome b5 stimulates purified testicular microsomal cytochrome P450 (C21 side-chain cleavage). Biochem Biophys Res Commun 1982;108:454–460.
24 Auchus RJ, Lee TC, Miller WL: Cytochrome b5 augments the 17,20-lyase activity of human P450c17 without direct electron transfer. J Biol Chem 1998; 273:3158–3165.
25 Conley AJ, Bird IM: The role of cytochrome P450 17α-hydroxylase and 3ß-hydroxysteroid dehydrogenase in the integration of gonadal and adrenal steroidogenesis via the Δ5 and Δ4 pathways of steroidogenesis in mammals. Biol Reprod 1997;56: 789–799.
26 Rainey WE: Adrenal zonation: clues from 11ß-hydroxylase and aldosterone synthase. Mol Cell Endocrinol 1999;151:151–160.
27 Auchus RJ: Overview of dehydroepiandrosterone biosynthesis. Semin Reprod Med 2004;22:281–288.
28 Lee-Robichaud P, Wright JN, Akhtar ME, Akhtar M: Modulation of the activity of human 17α-hydroxylase-17,20-lyase (CYP17) by cytochrome b5:endocrinological and mechanistic implications. Biochem J 1995;308:901–908.
29 Flück CE, Miller WL, Auchus RJ: The 17, 20-lyase activity of cytochrome P450c17 from human fetal testis favors the Δ5 steroidogenic pathway. J Clin Endocrinol Metab 2003;88:3762–3766.
30 Parker CR Jr: Dehydroepiandrosterone and dehydroepiandrosterone sulfate production in the human adrenal during development and aging. Steroids 1999;64:640–647.
31 Seron-Ferre M, Tayler NF, Rotten D, Koritnik DR, Jaffe RB: Changes in fetal rhesus monkey plasma dehydroepiandrosterone sulfate: relationship to gestational age, adrenal weight and preterm delivery. J Clin Endocrinol Metab 1983;57:1173–1178.
32 Kemnitz JW, Roecker EB, Haffa AL, Pinheiro J, Kurzman I, Ramsey JJ, MacEwen EG: Serum dehydroepiandrosterone sulfate concentrations across the life span of laboratory-housed rhesus monkeys. J Med Primatol 2000;29:330–337.
33 Koritnik DR, Laherty RF, Rotten D, Jaffe RB: A radioimmunoassay for dehydroepiandrosterone sulfate in the circulation of rhesus monkeys. Steroids 1983;42:653–667.
34 Levine J, Wolfe LG, Schiebinger RJ, Loriaux DL, Cutler GB Jr: Rapid regression of fetal adrenal zone and absence of adrenal reticular zone in the marmoset. Endocrinology 1982;111:1797–1802.
35 Pepe GJ, Waddell BJ, Albrecht ED: The effects of adrenocorticotropin and prolactin on adrenal dehydroepiandrosterone secretion in the baboon fetus. Endocrinology 1988;122:646–650.
36 Palmert MR, Hayden DL, Lansfield MJ, Crigler JF, Crowley WF, Chandler DW, Poepple PA: The longitudinal study of adrenal maturation during gonadal suppression: evidence that adrenarche is a gradual process. J Clin Endocrinol Metab 2001;86:4536–4542.
37 Martin DD, Schweizer R, Schwarze CP, Elmlinger MW, Ranke MB, Binder G: The early dehydroepiandrosterone sulfate rise of adrenarche and the delay of pubarche indicate primary ovarian failure in Turner syndrome. J Clin Endocrinol Metab 2004;89: 1164–1168.
38 Parker CR Jr, Carr BR, Simpson ER, MacDonald PC: Decline in the concentration of low-density lipoprotein-cholesterol in human fetal plasma near term. Metabolism 1983;32:919–923.
39 Reynolds J: Development and Function of the Human Fetal Adrenal Cortex. Fetal Endocrinology 1981;35–51.

40 Parker CR Jr, Carr BR, Winkel CA, Casey ML, Simpson ER, MacDonald PC: Hypercholesterolemia due to elevated low density lipoprotein-cholesterol in newborns with anencephaly and adrenal atrophy. J Clin Endocrinol Metab 1983;57:37–43.
41 Parker CR Jr, Atkinson MW, Owen J, Andrews WW: Dynamics of the fetal adrenal, cholesterol, and apolipoprotein B responses to antenatal betamethasone therapy. Am J Obstet Gynecol 1996;174: 562–565.
42 Jaffe RB, Seron-Ferre M, Huhtaniemi I, Korenbrot C: Regulation of the primate fetal adrenal gland and testis in vitro and in vivo. J Steroid Biochem 1977; 8:479–490.
43 Benner MC: Studies on the involution of the fetal cortex of the adrenal glands. Am J Pathol 1940;16: 787–798.
44 Abraham GE, Buster JE, Kyle FW, Corrales PC, Teller RC: Radioimmunoassay of plasma pregnenolone, 17-hydroxypregnenolone and dehydroepiandrosterone under various physiological conditions. J Clin Endocrinol Metab 1973;37:140–144.
45 Ducharme JR, Forret MG, de Perreti E, Sempe M, Collu R, Bertrand J: Plasma adrenal and gonadal sex steroids in human pubertal development. J Clin Endocrinol Metab 1976;42:468–476.
46 Azziz R, Farah LA, Moran C, Knochenhauer ES, Potter HD, Boots LR: Early adrenarche in normal prepubertal girls: a prospective longitudinal study. J Pediatr Endocrinol Metab 2004;17:1231–1237.
47 Remer T, Boye KR, Hartmann MF, Wudy SA: Urinary markers of adrenarche: reference values in healthy subjects, aged 3–18 years. J Clin Endocrinol Metab 2005;90:2015–2021.
48 Apter D, Pakarinen A, Hammond GL, Vihko R: Adrenocortical function in puberty. serum ACTH, cortisol and dehydroepiandrosterone in girls and boys. Acta Paediatr Scand 1979;68:599–604.
49 Mellon SH, Shively JE, Miller WL: Human proopiomelanocortin-(79–96), a proposed androgen stimulatory hormone, does not affect steroidogenesis in cultured human fetal adrenal cells. J Clin Endocrinol Metab 1991;72:19–22.
50 Parker LN, Lifrak ET, Odell WD: A 60,000 molecular weight human pituitary glycopeptide stimulates adrenal androgen secretion. Endocrinology 1983; 113:2092–2096.
51 Penhoat A, Sanchez P, Jaillard C, Langlois D, Begeot M, Saez JM: Human proopiomelanocortin-(79–96), a proposed cortical androgen-stimulating hormone, does not affect steroidogenesis in cultured human adult adrenal cells. J Clin Endocrinol Metab 1991; 72:23–26.
52 Adams JB: Control of secretion and the function of C19-delta 5-steroids of the human adrenal gland. Mol Cell Endocrinol 1985;41:1–17.
53 l'Allemand D, Schmidt S, Rousson V, Brabant G, Gasser T, Gruters A: Associations between body mass, leptin, IGF-I and circulating adrenal androgens in children with obesity and premature adrenarche. Eur J Endocrinol 2002;146:537–543.
54 Orentreich N, Brind JL, Rizer RL, Vogelman JH: Age changes and sex ifferences in serum dehydroepiandrosterone sulfate concentrations throughout adulthood. J Clin Endocrinol Metab 1984;59: 551–555.
55 Belanger A, Candas B, Dupont A, Cusan L, Diamond P, Gomez GL, Labrie F: Changes in serum concentrations of conjugated and unconjugated steroids in 40- to 80-year-old men. J Clin Endocrinol Metab 1994;79: 1086–1090.
56 Lasley BL, Santoro N, Randolf JF, Gold EB, Crawford S, Weiss G, McConnell DS, Sowers MF: The relationship of circulating dehydroepiandrosterone, testosterone, and estradiol to stages of the menopausal transition and ethnicity. J Clin Endocrinol Metab 2002;87:3760–3767.
57 Hornsby PJ: Biosynthesis of DHEAS by the human adrenal cortex and its age-related decline. Ann N Y Acad Sci 1995;774:29–46.
58 Parker CR Jr, Mixon RL, Brissie RM, Grizzle WE: Aging alters zonation in the adrenal cortex of men. J Clin Endocrinol Metab 1997;82:3898–3901.
59 Dharia S, Slane A, Jian M, Conner M, Conley AJ, Brissie RM, Parker CR Jr: Effects of aging on cytochrome b5 expression in the human adrenal gland. J Clin Endocrinol Metab 2005;90:4357–4361.
60 Hopper B, Tullner WW: Urinary estrone and plasma progesterone levels during the menstrual cycle of the rhesus monkey. Endocrinology 1970;86: 1225–1230.
61 Walsh SW, Resko JA, Grumbach MM, Novy MJ: In utero evidence for a functional fetoplacental unit in rhesus monkeys. Biol Reprod 1980;23:264–270.
62 Walsh SW, Norman RL, Novy MJ: In utero regulation of rhesus monkey fetal adrenals: effects of dexamethasone, adrenocorticotropin, thyrotropin-releasing hormone, prolactin, human chorionic gonadotropin, and alpha-melanocyte-stimulating hormone on fetal and maternal plasma steroids. Endocrinology 1979; 104:1805–1813.
63 Hum DW, Belanger A, Levesque E, Barbier O, Beaulieu M, Albert C, Vallee M, Guillemette C, Tchernof A, Turgeon D, Dubois C: Characterization of UDP-glucuronosyltransferases active on steroid hormones. J Steroid Biochem Mol Biol 1999;69: 413–423.
64 Gorwill RH, Snyder DL, Lindholm UB, Jaffe RB: Metabolism of pregnenolone-4-14C and pregnenolone-7-a-3H sulfate by the *Macaca mulatta* fetal adrenal in vitro. Gen Comp Endocrinol 1971;16: 21–29.

65 Snyder DL, Goebelsmanen U, Jaffe RB, Kirton KT: Extent of in vivo aromatization of dehydroepiandrosterone sulfate and dehydroepiandrosterone by the perfused Macaca mulatta placenta. Endocrinology 1971;88:274–278.
66 Walsh SW, Stanczyk FZ, Novy MJ: Daily hormonal changes in the maternal, fetal, and amniotic fluid compartments before parturition in a primate species. J Clin Endocrinol Metab 1984;58:629–639.
67 Kittinger GW, Beamer NB, Hagemenas F, Hill JD, Baughmann WL, Ochsner AJ: Evidence for autonomous pituitary-adrenal function in the near-term fetal rhesus (*Macaca mulatta*). Endocrinology 1972; 91:1037–1044.
68 Coulter CL, Read RC, Barry SJ, Tarantal AF, Styne DF: Role of hypothalamic-pituitary axis in EGF action on maturation of adrenal gland in fetal rhesus monkey in vivo. Pediatr Res 2001;50:210–216.
69 Seron-Ferre M, Hess DL, Lindholm U, Jaffe RB: Persistence of fetal zone function in the infant rhesus monkey adrenal gland. J Clin Endocrinol Metab 1986;62:460–465.
70 Jaffe RB, Seron-Ferre M, Parer JT, Lawrence CC: The primate fetal pituitary-adrenal axis in the perinatal period. Am J Obstet Gynecol 1978;131:164–170.
71 Arlt W, Martens JW, Song M, Wang JT, Auchus RJ, Miller WL: Molecular evolution of adrenarche: structural and functional analysis of P450c17 from four primate species. Endocrinology 2002;143:4665–4672.
72 Pepe GJ, Albrecht ED: Central integrative role of oestrogen in the regulation of placental steroidogenic maturation and the development of the fetal pituitary-adrenocortical axis in the baboon. Hum Reprod Update 1998;4:406–419.
73 Dawood MY, Fuchs F: Estradiol and progesterone in the maternal and fetal circulation in the baboon. Biol Reprod 1980;22:179–184.
74 Albrecht ED, Haskins AL, Pepe GJ: The influence of fetectomy at midgestation upon the serum concentrations of progesterone, estrone, and estradiol in baboons. Endocrinology 1980;107:766–770.
75 Ducsay CA, Hess DL, McClellan MC, Novy MJ: Endocrine and morphological maturation of the fetal and neonatal adrenal cortex in baboons. J Clin Endocrinol Metab 1991;73:385–395.
76 Albrecht ED, Pepe GJ: Effect of estrogen on dehydroepiandrosterone formation by baboon fetal adrenal cells in vitro. Am J Obstet Gynecol 1987; 156:1275–1278.
77 Albrecht ED, Aberdeen GW, Pepe GJ: Estrogen elicits cortical zone-specific effects on development of the primate fetal adrenal gland. Endocrinology 2005;146:1737–1744.
78 Albrecht ED, Babischkins JS, Davis WA, Leavitt MG, Pepe GJ: Identification and developmental expression of the estrogen receptor alpha and beta in the baboon fetal adrenal gland. Endocrinology 1999;140:5953–5961.
79 Abbott DH, Barnett DK, Colman RJ, Yamamoto ME, Schult-Darken NJ: Aspects of common marmoset basic biology and life history important for biomedical research. Comp Med 2003;53:339–350.
80 Pattison JC, Abbott DH, Saltzmann W, Nguyen AD, Henderson G, Jing H, Pryce CR, Allen AJ, Conley AJ, Bird IM: Male marmoset monkeys express an adrenal fetal zone at birth, but not a zona reticularis in adulthood. Endocrinology 2005;146:365–374.
81 Hodges JK, Brand H, Henderson C, Kelly RW: Levels of circulating and urinary oestrogens during pregnancy in the marmoset monkey *(Callithrix jacchus)*. J Reprod Fertil 1983;67:73–82.
82 Chambers PL, Hearn JP: Peripheral plasma levels of progesterone, oestradiol-17ß, oestrone, testosterone, androstenedione and chorionic gonadotrophin during pregnancy in the marmoset monkey, *Callithrix jacchus*. J Reprod Fertil 1979;56:23–32.
83 Pattison JC, Saltzmann W, Abbott DH, Hogan BK, Nguyen AD, Husen B, Einspanier A, Conley AJ, Bird IM: Gender and gonadal status differences in zona reticularis expression in marmoset monkey adrenals: Cytochrome b5 localization with respect to cytochrome P450 17,20-lyase activity. Mol Cell Endocrinol 2007;265–266:93–101.
84 Hill W: Observations on the growth of the suprarenal cortex. J Anatomy 1930;64:479–502.
85 Scammon RE: The prenatal growth and natal involution of the human suprarenal gland. Proc Soc Exp Biol Med 1926;26:809–811.
86 Lanman J: The Fetal Zone of the Adrenal Gland. Medicine 1953;32:389–429.
87 Grollman A: The Adrenals. Williams & Wilkins Company, Baltimore, 1936.
88 Sucheson M: Development of Zonular Patterns in the Human Adrenal Gland. J Morphol 1968;126:477–492.
89 Young MC, Laurence KM, Hughes IA: Relationship between fetal adrenal morphology and anterior pituitary function. Horm Res 1989;32:130–135.
90 Seron-Ferre M, Lawrence CC, Siiteri PK, Jaffe RB: Steroid production by definitive and fetal zones of the human fetal adrenal gland. J Clin Endocrinol Metab 1978;47:603–609.
91 Sakhatskaya TS, Altukhova VI: Formation of dehydroepiandrosterone sulfate and hydrocortisone in the definitive and fetal adrenal cortex of human fetuses. Sov J Dev Biol 1973;4:46–50.

92 Nelson HP, Kuhn RW, Deyman ME, Jaffe RB: Human fetal adrenal definitive and fetal zone metabolism of pregnenolone and corticosterone: alternate biosynthetic pathways and absence of detectable aldosterone synthesis. J Clin Endocrinol Metab 1990;70:693–698.
93 Mesiano S, Coulter CL, Jaffe RB: Localization of cytochrome P450 cholesterol side-chain cleavage, cytochrome P450 17α-hydroxylase/17, 20-lyase, and 3ß-hydroxysteroid dehydrogenase isomerase steroidogenic enzymes in human and rhesus monkey fetal adrenal glands: reappraisal of functional zonation. J Clin Endocrinol Metab 1993;77:1184–1189.
94 Narasaka T, Suzuki T, Moriya T, Sasano H: Temporal and spatial distribution of corticosteroidogenic enzymes immunoreactivity in developing human adrenal. Mol Cell Endocrinol 2001;174:111–120.
95 Dharia S, Slane A, Jian M, Conner M, Conley AJ, Parker CR Jr: Colocalization of P450c17 and cytochrome b5 in androgen-synthesizing tissues of the human. Biol Reprod 2004;71:83–88.
96 Dupont E, Luu-The V, Labrie F, Pelletier G: Ontogeny of 3ß-hydroxysteroid dehydrogenase/ Δ5–4 isomerase (3ß-HSD) in human adrenal gland performed by immunocytochemistry. Mol Cell Endocrinol 1990;74:R7–R10.
97 Rehman KS, Carr BR, Rainey WE: Profiling the steroidogenic pathway in human fetal and adult adrenals. J Soc Gynecol Investig 2003;10:372–380.
98 Coulter CL, Jaffe RB: Functional maturation of the primate fetal adrenal in vivo: 3. Specific zonal localization and developmental regulation of CYP21A2 (P450c21) and CYP11B1/CYP11B2 (P450c11/aldosterone synthase) lead to integrated concept of zonal and temporal steroid biosynthesis. Endocrinology 1998;139:5144–5150.
99 Hornsby PJ, Aldern KA: Steroidogenic enzyme activities in cultured human definitive zone adrenocortical cells: comparison with bovine adrenocortical cells and resultant differences in adrenal androgen synthesis. J Clin Endocrinol Metab 1984;58:121–127.
100 Mapes S, Tarantal AF, Parker CR, Moran FM, Bahr JM, Pyter L, Conley AJ: Adrenocortical cytochrome b5 expression during fetal development of the rhesus macaque. Endocrinology 2002;143:1451–1458.
101 Miraglia T, Moreira E: The Adrenal Cortex of the Marmoset. Acta Anat 1969;72:520–532.
102 Leavitt MG, Albrecht ED, Pepe GJ: Development of the baboon fetal adrenal gland: regulation of the ontogenesis of the definitive and transitional zones by adrenocorticotropin. J Clin Endocrinol Metab 1999;84:3831–3835.
103 Coulter CL, Goldsmith PC, Mesiano S, Voytek CC, Martin MC, Mason JI, Jaffe RB: Functional maturation of the primate fetal adrenal in vivo. II. Ontogeny of corticosteroid synthesis is dependent upon specific zonal expression of 3ß-hydroxysteroid dehydrogenase/isomerase. Endocrinology 1996;137:4953–4959.
104 McNulty WP, Novy MJ, Walsh SW: Fetal and postnatal development of the adrenal glands in *Macaca mulatta*. Biol Reprod 1981;25:1079–1089.
105 Parker CR Jr, Jian M, Conley AJ: The localization of DHEA sulfotransferase in steroidogenic and steroid metabolizing tissues of the adult rhesus macaque monkey. Endocr Res 2000;26:517–522.
106 Mapes S, Corbin CJ, Tarantal AF, Conley AJ: The primate adrenal zona reticularis is defined by expression of cytochrome b5, 17α-hydroxylase/17,20-lyase cytochrome P450 (P450c17) and NADPH-cytochrome P450 reductase (reductase) but not 3ß-hydroxysteroid dehydrogenase/Δ5–4 isomerase (3ß-HSD). J Clin Endocrinol Metab 1999;84:3382–3385.

Alan J. Conley
Department of Population Health and Reproduction
School of Veterinary Medicine, University of California, Davis
1114 Tupper Hall, Davis, CA 95616 (USA)
Tel. +1 530 752 2128, Fax +1 530 752 4278, E-Mail ajconley@ucdavis.edu

Flück CE, Miller WL (eds): Disorders of the Human Adrenal Cortex.
Endocr Dev. Basel, Karger, 2008, vol 13, pp 55–66

Clinical Implications of Androgen Synthesis via 5α-Reduced Precursors

Hans K. Ghayee · Richard J. Auchus

Division of Endocrinology and Metabolism, Department of Internal Medicine,
University of Texas Southwestern Medical Center, Dallas, Tex., USA

Abstract

During embryogenesis, the male external genitalia are formed by the action of the potent androgen, dihydrotestosterone (DHT). DHT is produced in human genital skin and prostate from testosterone via the action of 5α-reductase type 2. The biological relevance of this pathway to DHT is evidenced by patients with mutations in the gene encoding 5α-reductase type 2, which causes severely undermasculinized external genitalia in genetic males. In contrast, this paradigm of androgen physiology does not explain some clinical observations, such as the differences noted in the virilization of females with various congenital adrenal hyperplasias. An alternate pathway to DHT was elucidated in the tammar wallaby pouch young, and studies in knockout mice showed that this pathway uses 5α-reductase type 1 to convert 17-hydroxyprogesterone to 5α-reduced androgen precursors. Flux via the alternate or 'backdoor' pathway has been implicated in human diseases such as P450 oxidoreductase deficiency, polycystic ovarian disease, and congenital adrenal hyperplasia. A better understanding of the 5α-reduced or 'backdoor' pathway to DHT in human disorders of androgen excess will provide pharmacotherapy opportunities to effectively treat androgen excess in females.

The 5α-Reduction of Testosterone to Dihydrotestosterone

Among lay audiences, testosterone (T) is considered the 'male hormone', to contrast with estradiol, which is colloquially the 'female hormone'. This dichotomy is seemingly justified because T is the principal secretion of the adult human testis, whereas estradiol is the biologically relevant estrogen. The true 'male hormone', however, is not T but its 5α-reduced metabolite, dihydrotestosterone (DHT) [1]. DHT is responsible for the formation of the external genitalia (labioscrotal fusion, scrotum formation, and penile growth) in the male fetus and for most aspects of sexual maturation at puberty. Consequently, DHT is the hormone that mediates most of the biologic changes which phenotypically distinguish males from females.

Bruchovsky and Wilson [2] discovered the enzymatic conversion of T to DHT in rat prostate 40 years ago. Subsequently, perineoscrotal pseudovaginal hypospadias was described, a disorder in which genetic males were born with severely undermasculinized external genitalia and often were raised as females. Wilson's group demonstrated that the defect in this disorder was a deficiency of one of the two 5α-reductase enzymes in genital skin [3, 4], and molecular genetic analyses two decades later showed that the gene for 5α-reductase type 2 *(SRD5A2)* was mutated in this disease [5]. Thus, a paradigm was firmly established, that the 5α-reduction of T is the biologically relevant pathway of DHT synthesis in human beings.

This paradigm derives from the classical pathways of androgen biosynthesis. In human beings, the dominant route to 19-carbon steroids in the adrenal [6] and testis [7] is from pregnenolone to dehydroepiandrosterone and its sulfate (DHEAS), which are sequentially converted to androstenedione, T, and DHT in the gland or in peripheral tissues (fig. 1). Like most simple models in biology, however, this paradigm fails to explain all aspects of androgen physiology. For example, female fetuses with 21-hydroxylase deficiency (21OHD) can virilize severely [8], due to the production of 19-carbon steroids derived from the adrenal gland, presumably DHEAS. In contrast, newborn girls with 3β-hydroxysteroid dehydrogenase type 2 deficiency (3βHSDD) show little labioscrotal fusion and only moderate clitoral enlargement [9], yet the fetal adrenal in 3βHSDD produces as much or more DHEAS as in 21OHD [10]. Because DHT is the androgen that causes virilization of the external genitalia and because the peripheral pathways from DHEAS to DHT are normal in both disorders, one would anticipate equivalent virilization of females in 3βHSDD and 21OHD if adrenal DHEAS was the relevant adrenal precursor to DHT in 21OHD.

In addition, T is neither the only nor the best substrate for the human 5α-reductases, particularly the type 1 enzyme. Virtually all 3-keto-Δ^4-steroids are metabolized by both 5α- and 5β-reduction prior to being excreted in the urine. In fact, the 21-carbon steroids progesterone, 17-hydroxyprogesterone, and 4-pregnene-17α,20α-diol-3-one are better substrates for the human 5α-reductases than T, particularly for the type 1 isoenzyme [11]. The 5α-reduction of 21-carbon steroids has been considered largely catabolic reactions mediated by the type 1 isoenzyme in the liver, with some exceptions such as allopregnanolone formation in the brain [12]. Could these 5α-reduced progesterone derivatives also serve as precursors for DHT?

Dihydrotestosterone Formation in the Tammar Wallaby Pouch Young

Marsupials have been used for many years as a model for the study of sexual differentiation in mammals [13]. The pouch young or joey is born in a state where the inferior half of the body is completely undeveloped, including the external genitalia. The joey climbs into the pouch using its forelimbs and attaches to a nipple, and sexual differentiation occurs outside the uterus. The marsupial model therefore provides an experimentally approachable system in which one may administer steroids or drugs, measure circulating concentrations of steroids, and obtain tissue specimens [14]. In

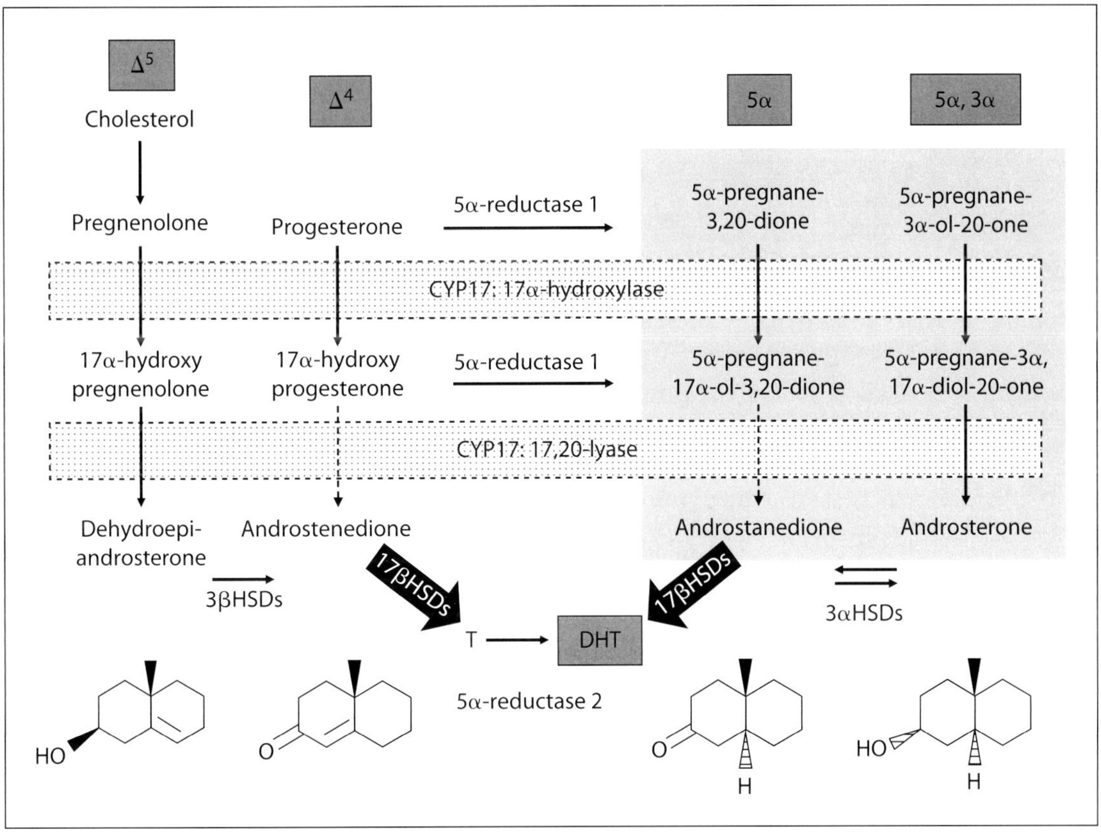

Fig. 1. The four possible pathways from 21-carbon precursors to 19-carbon steroids. The A/B-ring structures are shown at the bottom for the Δ^5, Δ^4, 5α-reduced, and 5α-, 3α-reduced steroids (labeled at top). The steroid names and enzymes are indicated, with the 5α-reduced and 5α-, 3α-reduced steroids in the shaded area. Human CYP17A1 17α-hydroxylates all four classes of 21-carbon steroids well, but the 17,20-lyase activity is only efficient for the Δ^5- and 5α-, 3α-pathways (solid versus dashed lines).

the tammar wallaby, however, T concentrations are low and without sexually dimorphic patterns during the time of sexual differentiation [15]. These data suggested that T was not the circulating androgen responsible for formation of the external genitalia in this species and that the 5α-reduction of T was not the relevant pathway to DHT in the developing male pouch young.

Further studies demonstrated that 5α-androstane-3α,17β-diol (Adiol) was the androgen that circulates with a sexually dimorphic profile during sexual differentiation in the tammar wallaby and enabled prostate formation [16]. Although Adiol can be a catabolic metabolite of DHT, Adiol is also a potent androgen precursor and is superior to T or DHT in the pharmacologic induction of prostate hyperplasia in dogs [17], since Adiol is readily converted to DHT in the prostate by an oxidative 3α-HSD

[18]. The observation that Adiol demonstrated sexually dimorphic circulating concentrations in the tammar wallaby pouch young raised the question of whether Adiol was not primarily a metabolite of DHT but rather an obligate precursor of DHT in this species.

To elucidate the pathway of Adiol synthesis in the tammar wallaby pouch young, testis were removed from animals of the appropriate developmental age and incubated with radiolabeled steroid precursors with and without the 5α-reductase inhibitor 17β-(N,N-diethyl)carbamoyl-4-methyl-4-aza-5α-androstan-3-one (4-MA). The time course of steroid products enabled us to deduce the pathway of Adiol synthesis from progesterone. The first surprising result was that, after progesterone was converted to 17-hydroxyprogesterone, the expected metabolites androstenedione, T, and DHT were not observed. Instead, 17-hydroxyprogesterone was converted into an unknown compound, whose formation was blocked by 4-MA [19]. This experiment was the first clue that Adiol synthesis involved the 5α-reduction of a 21-carbon steroid precursor, namely 17-hydroxyprogesterone.

Additional experiments showed that the immediate product of 17-hydroxyprogesterone reduction, 5α-pregnane-17α-ol-3,20-dione, did not have the same chromatographic mobility as the unknown product. We recognized that most 5α-reduced steroids are good substrates for reductive 3α-HSDs, enzymes of the aldo-keto reductase family that are ubiquitously expressed in tissues [20], and that the ultimate product Adiol was both 5α- and 3α-reduced. Indeed, the unknown compound was 5α-pregnane-3α,17α-diol-20-one (Pdiol), which is 17-hydroxyprogesterone after both 5α- and 3α-reduction. Subsequent experiments confirmed that Pdiol was cleaved to the 19-carbon steroid androsterone, which served as the immediate precursor to Adiol [19]. Adiol was converted in the prostate to DHT, completing this alternate pathway to DHT [16] (fig. 2).

As is often true in science, an alternate pathway to Adiol was suggested by earlier experiments in other systems. Using immature rat testes, trapping experiments suggested that Adiol was synthesized by a pathway in which androstenedione and T were not intermediates [21]. However, the relevance of this observation was not clear, since the mass of Adiol made is small and insufficient to induce pubertal maturation. In addition, Adiol is synthesized by neonatal testes of several species [22–26] and by adult rodent testes after administration of long-acting GnRH agonists [27].

We found that androgen biosynthesis in the fetal mouse testis is brisk and primarily via the conventional pathway, progesterone to 17-hydroxyprogesterone to androstenedione to T [28]. There was no evidence of significant 5α-reductase activity in the fetal mouse testis. In contrast, we also found that Adiol was produced by the immature mouse testes (postnatal day 25), although the major products of progesterone metabolism were the 5α-reduced 21-carbon steroids 5α-pregnane-3,20-dione and 5α-pregnane-3α-ol-20-one (allopregnanolone). Adiol was the major 19-carbon steroid product, but time course experiments showed that the dominant route to Adiol was via the conventional pathway (not the wallaby pathway) of 17-hydroxyprogesterone

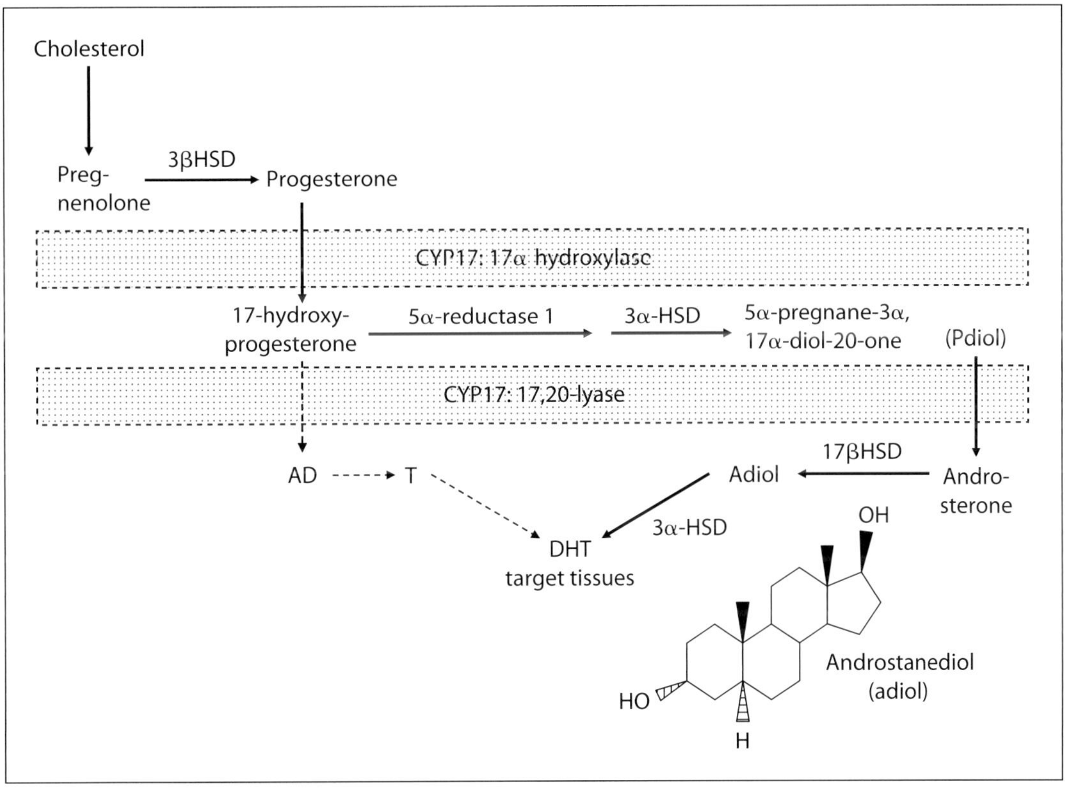

Fig. 2. The alternate pathway to DHT in the tammar wallaby. Progesterone is converted to 17-hydroxyprogesterone, then both 5α- and 3α-reduced to Pdiol. Pdiol is cleaved to androsterone, which is reduced to Adiol, whose structure is shown at bottom right.

to androstenedione to T to DHT and then to Adiol. Using mice with targeted deletion of the 5α-reductase genes, we showed that the isoenzyme in the neonatal mouse testis is the type 1 [28].

The Alternate Pathway to Dihydrotestosterone in Human Disease

Although this 5α-reduced pathway is the biologically relevant route of DHT synthesis in the tammar wallaby pouch young, it is unlikely that this pathway contributes significantly to the formation of the external genitalia in the male human fetus. The observation that genetic defects in 5α-reductase type 2 are sufficient to prevent genital masculinization in males [29], the absence of the type 2 isoenzyme in the testis, and the high T concentrations in patients with *SRD5A2* mutations, and T synthesis in isolated fetal testis [30] all suggest that T is the principal circulating androgen during human development. If this pathway is relevant at all, then the human enzyme cytochrome P450c17 (CYP17A1) must be capable of converting Pdiol to androsterone. We asked this question using recombinant human CYP17A1 expressed in yeast microsomes with human P450-oxidoreductase (POR) [6]. We showed that both 5α-reduced 21-carbon steroids 5α-pregnane-3,20-dione and 5α-pregnane-3α-ol-20-one are substrates for the 17α-hydroxylase activity of human CYP17A1 [31], the latter substrate yielding Pdiol. The 5α-pregnane-3,20-dione metabolism stops after 17-hydroxylation without significant cleavage to 19-carbon steroids by the 17,20-lyase activity of CYP17A1. In contrast, Pdiol is an excellent substrate for the 17,20-lyase reaction, yielding androsterone as the 19-carbon steroid product [31]. Pdiol is the best substrate yet found for the 17,20-lyase activity of human CYP17A1, with a K_m comparable to that of the conventional 'best substrate' 17α-hydroxypregnenolone and a 10-fold higher V_{max}.

The 17,20-lyase activity of CYP17A1 requires the cofactor protein cytochrome b5 for optimal activity, and cytochrome b5 increases the rate of 17α-hydroxypregnenolone cleavage to DHEA by a factor of 10 [6, 32, 33]. Consequently, we compared the 17,20-lyase activity using these substrates and recombinant human CYP17A1 with or without cytochrome b5, and we found that cytochrome b5 augments the conversion of Pdiol to androsterone by only a factor of 3. Nonetheless, the rate of Pdiol cleavage to androsterone without cytochrome b5 is still faster than the conversion of 17-hydroxypregnenolone to DHEA in the presence of cytochrome b5 [31]. Consequently, the human enzymes are capable of using the 5α-reduced pathway to DHT. In particular, this pathway will be functional even in tissues that lack cytochrome b5, including the zona fasciculata of the adrenal cortex and possibly the immature testis.

If the alternate or 'backdoor pathway' [34] to DHT is functional in human beings, then it is probably most relevant in pathologic states and probably in the female. The three requirements for significant flux via the 5α-reduced pathway include the presence of 5α-reductase, the presence of CYP17A1, and high concentrations of 17OHP. The expression of 5α-reductase in human steroidogenic tissue is best described for the adult ovary. In the luteal phase, comparable amounts of 5α-pregnane-3,20-dione are produced with progesterone [35], and studies of 46,XX women with 5α-reductase type 2 deficiency indicate that the ovarian 5α-reductase is the type 1 isoenzyme [36]. In the rat, 5α-reductase type 1 is expressed in the zona fasciculata, and its abundance is attenuated by androgens and increased with castration [37]. The human adrenal cortex does not appear to contain much 5α-reductase activity, but the expression of 5α-reductases in the human fetal adrenal gland has not been studied adequately [38]. Nevertheless, have clinical conditions been described in which 17-hydroxyprogesterone accumulates, androgen excess is observed, and 5α-reduced androgens are dominant?

During the time that we were elucidating the biochemistry of the 5α-reduced pathway to DHT, a new disorder of steroidogenesis was being characterized with molecular genetics. In 1985, children with apparent deficiencies in both 17- and 21-hydroxylase activities were described [39], based on the accumulation of both corticosterone and

17-hydroxyprogesterone, respectively [40]. The genetic basis of this disease remained elusive until the completion of the Human Genome Project allowed Miller and colleagues to explore their proposal [41] that these patients had deficiencies in POR. These patients were found to have mutations in the *POR* gene that partially and variably impair the capacity of the POR flavoprotein to transfer electrons to various microsomal P450s [42–44]. The production of 19-carbon steroids is severely reduced in these patients, which reflects the vulnerability of the 17,20-lyase reaction to disruptions in interactions with POR [45–47]. Although the low 17,20-lyase activity postnatally is consistent with the incomplete genital virilization of males with POR deficiency, the external genitalia of girls with this same disorder are partially masculinized at birth. Given that 17-hydroxyprogesterone accumulates in POR deficiency, the paradox of low androgen synthesis postnatally and virilization of female infants in utero might be reconciled by the alternative pathway to DHT in the fetal adrenal gland.

During a normal pregnancy, the fetal adrenal produces massive amounts of DHEAS, which is desulfated, hydroxylated, oxidized, isomerized, aromatized, and reduced to estriol, the estrogen most distinctly elevated during pregnancy. Estriol production is very low during pregnancies of fetuses with diseases that affect any step in the pathway, including POR deficiency. Nevertheless, the pattern of steroid metabolites found in urine of pregnant women carrying female POR-deficient fetuses reflects the compounds produced by the fetal adrenal glands. In a normal pregnancy, the major 19-carbon steroid produced by the fetus is DHEAS, which is converted to androstenedione and then either 5α-reduced or 5β-reduced. Both products are efficiently 3α-reduced, yielding roughly a 1:1 ratio of androsterone and etiocholanolone (fig. 3), and these reduced steroids are glucuronidated and excreted in the urine. In the POR-deficient adrenal gland, the poor 17,20-lyase activity and high 17-hydroxyprogesterone might drive a small amount of steroidogenesis via the alternate pathway. If the fetal adrenal did produce 5α-reduced 19-carbon steroids directly, this process would be reflected by a high ratio of 5α-/5β-reduced steroids in the urine and a high androsterone/etiocholanolone ratio (fig. 3). Consistent with this proposal, the urinary androsterone/etiolcholanolone ratio in these pregnancies is elevated 4-fold, and Pdiol is demonstrable in the urine as well [48]. Androsterone is also disproportionately elevated in the urine of infants with POR deficiency [49], including girls, which further supports that flux via the alternate pathway occurs in POR deficiency. Note that, unlike DHEAS, even small amounts of androsterone are capable of partially virilizing a female fetus, since androsterone is already 5α-reduced and an excellent precursor of DHT. It is also possible that androsterone from the fetal adrenal limits the undermasculinization of boys with POR deficiency, given that the reduction in T production to this degree ordinarily causes severe genital ambiguity.

As suggested above, another more common disorder in which 17OHP accumulates in the fetal adrenal is 21OHD, and the discrepancy in virilization of female infants born with 21OHD and 3βHSDD is not explained by the conventional pathway to DHEAS, followed by conversion to T and 5α-reduction to DHT. In considering

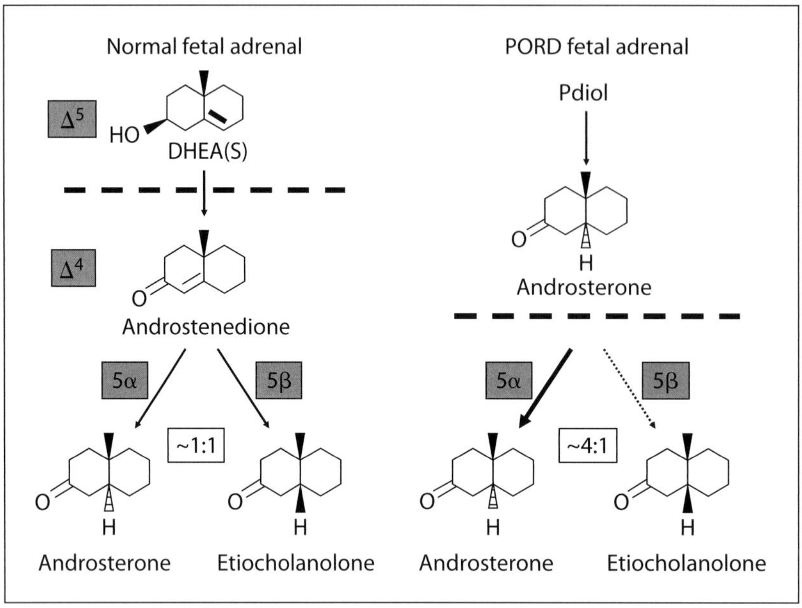

Fig. 3. Androsterone and etiocholanolone production via adrenal CYP17A1 in the normal adrenal and in POR deficiency (PORD). The normal adrenal (left) produces DHEAS, which is metabolized in the periphery to androstenedione, a Δ^4-steroid. Since Δ^4-steroids are good substrates for both 5α- and 5β-reductases, a roughly 1:1 ratio of the two 19-carbon steroids is excreted in the urine. In PORD, the androsterone/etiocholanolone ratio in the urine is elevated in the fetus and neonate, suggesting that the adrenal directly produces the 5α-reduced steroid androsterone, which cannot be 5β-reduced in peripheral tissues.

the alternate pathway to DHT, a key difference in steroid metabolism between these two conditions is an inability to make Δ^4-steroids in 3βHSD2-deficient adrenals, which precludes the 5α-reduction of precursors within the adrenal itself. In contrast, high intra-adrenal 17-hydroxyprogesterone in 21OHD might allow formation of Pdiol and androsterone via the alternate pathway. Experimental confirmation of adrenal Pdiol and androsterone production in 21OHD is incomplete because the sum of androsterone plus etiocholanolone is used diagnostically. Androsterone glucuronide rises briskly with cosyntropin stimulation in women with nonclassical 21OHD [50], consistent with its synthesis via the alternate pathway. Interindividual variation in the fraction of steroid precursors metabolized along the conventional and alternate pathways may explain some of the phenotypic variation observed in 21OHD patients with the same *CYP21A2* mutations.

Even more common than 21OHD is polycystic ovary syndrome (PCOS), a heterogeneous disorder that afflicts about 5% of reproductive-age women [51] and bears as its hallmark androgen excess. In normal fertile women, the majority of T, and presumably DHT, arises from peripheral conversion of adrenal DHEAS [52]. In PCOS, both ovarian and adrenal androgens are produced in excess, and ovarian androgen excess is a prominent feature. As a consequence of androgen synthesis, the ovary also produces 17-hydroxyprogesterone, and a portion of circulating 17-hydroxyprogesterone derives from the ovaries. Serum concentrations of 17-hydroxyprogesterone rise upon administration of a GnRH agonist [53], and this rise is exaggerated in women with functional ovarian hyperandrogenism [54] and PCOS [55]. This exaggerated 17-hydroxyprogesterone/androstenedione ratio has been interpreted as indicative of poor ovarian 17,20-lyase activity [54, 55] but may simply reflect high flux from precursors and the inability of human CYP17A1 to cleave 17-hydroxyprogesterone efficiently [6].

What happens to all that 17-hydroxyprogesterone? One possibility is that a portion is 5α-reduced before being cleaved to androsterone via the alternate pathway. The 5α-reduced 19-carbon steroids androsterone glucuronide [56] and androstanediol glucuronide [57] are elevated in hirsute women, perhaps more consistently than increased production of T, and at least some androsterone glucuronide is of adrenal origin [56]. In addition, peripheral 5α-reductase activity is increased in PCOS, but this same study showed increased urinary androsterone/etiocholanolone ratios in PCOS women versus normal controls [58]. Consequently, it is likely that a portion of the DHT produced in PCOS derives via the alternate pathway, possibly more so than in other women. If this is true, then additional enzymes along the alternate pathway might be good targets for therapeutic intervention in androgen-dependent disorders.

Future Directions

The relevance of the 5α-reduced or 'backdoor' pathway to DHT in human disorders of androgen excess in females is suggested by the biochemical evidence and supported by preliminary clinical studies. More detailed efforts designed to specifically dissect the proportion of steroids derived from the conventional and alternate pathways are necessary to determine exactly how 19-carbon steroids are synthesized in diseases that cause hirsutism and virilization in females. Genetic differences in the enzymes specific for this alternate pathway may explain the discrepancies between circulating T concentrations and clinical manifestations in women with androgen excess. Pharmacotherapy targeting these specific enzymes may provide effective therapy in androgen excess states and avoid some of the untoward effects of current modalities.

Acknowledgements

We thank our colleagues in Dallas and Australia who work with us on these studies, particularly Dr. Jean Wilson, who is the father of 5α-reductase and continues to lead us into new frontiers of androgen physiology. This work was supported by grant I-1493 from the Robert A. Welch Foundation to R.J.A.

References

1 Griffin JE, Wilson JD: Disorders of the testes and the male reproductive tract; in Larson PR, et al. (eds): Williams Textbook of Endocrinology, W.B. Saunders, 2003, pp 709–770.

2 Bruchovsky N, Wilson JD: The intranuclear binding of testosterone and 5α-androstan-17β-ol-3-one by rat prostate. J Biol Chem 1968;243:5953–5960.

3 Walsh PC, Madden JD, Harrod MJ, Goldstein JL, MacDonald PC, Wilson JD: Familial incomplete male pseudohermaphroditism, type 2. Decreased dihydrotestosterone formation in pseudovaginal perineoscrotal hypospadias. N Engl J Med 1974;291: 944–949.

4 Moore RJ, Wilson JD: Steroid 5α-reductase in cultured human fibroblasts. Biochemical and genetic evidence for two distinct enzyme activities. J Biol Chem 1976;251:5895–5900.

5 Andersson S, Berman DM, Jenkins EP, Russell DW: Deletion of a steroid 5α-reductase 2 gene in male pseudohermaphroditism. Nature 1991;354:159–161.

6 Auchus RJ, Lee TC, Miller WL: Cytochrome b_5 augments the 17,20 lyase activity of human P450c17 without direct electron transfer. J Biol Chem 1998; 273:3158–3165.

7 Flück CE, Miller WL, Auchus RJ: The 17, 20-lyase activity of cytochrome P450c17 from human fetal testis favors the Δ^5 steroidogenic pathway. J Clin Endocrinol Metab 2003;88:3762–3766.

8 White PC, Speiser PW: Congenital adrenal hyperplasia due to 21-hydroxylase deficiency. Endocr Rev 2000;21:245–291.

9 Moisan AM, Ricketts ML, Tardy V, Desrochers M, Mebarki F, Chaussain JL, Cabrol S, Raux-Demay MC, Forest MG, Sippell WG, Peter M, Morel Y, Simard J: New insight into the molecular basis of 3β-hydroxysteroid dehydrogenase deficiency: identification of eight mutations in the HSD3B2 gene in eleven patients from seven new families and comparison of the functional properties of twenty-five mutant enzymes. J Clin Endocrinol Metab 1999;84: 4410–4425.

10 Bongiovanni AM: The adrenogenital syndrome with deficiency of 3β-hydroxysteroid dehydrogenase. J Clin Invest 1962;41:2086.

11 Frederiksen DW, Wilson JD: Partial characterization of the nuclear reduced nicotinamide adenine dinucleotide phosphate: Δ^4-3-ketosteroid 5α-oxidoreductase of rat prostate. J Biol Chem 1971;246: 2584–2593.

12 Bitran D, Shiekh M, McLeod M: Anxiolytic effect of progesterone is mediated by the neurosteroid allopregnanolone at brain $GABA_A$ receptors. J Neuroendocrinol 1995;7:171–177.

13 Wilson JD, Shaw G, Leihy ML, Renfree MB: The marsupial model for male phenotypic development. Trends Endocrinol Metab 2002;13:78–83.

14 Wilson JD, Leihy MW, Shaw G, Renfree MB: Unsolved problems in male physiology: studies in a marsupial. Mol Cell Endocrinol 2003;211:33–36.

15 Wilson JD, George FW, Shaw G, Renfree MB: Virilization of the male pouch young of the tammar wallaby does not appear to be mediated by plasma testosterone or dihydrotestosterone. Biol Reprod 1999;61:471–475.

16 Shaw G, Renfree MB, Leihy MW, Shackleton CH, Roitman E, Wilson JD: Prostate formation in a marsupial is mediated by the testicular androgen 5α-androstane-3α,17β-diol. Proc Natl Acad Sci USA 2000;97:12256–12259.

17 Walsh PC, Wilson JD: The induction of prostatic hypertrophy in the dog with androstanediol. J Clin Invest 1976;1093–1097.

18 Biswas MG, Russell DW: Expression cloning and characterization of oxidative 17β- and 3α-hydroxysteroid dehydrogenases from rat and human prostate. J Biol Chem 1997;272:15959–15966.

19 Wilson JD, Auchus RJ, Leihy MW, Guryev OL, Estabrook RW, Osborn SM, Shaw G, Renfree MB: 5α-androstane-3α,17β-diol is formed in tammar wallaby pouch young testes by a pathway involving 5α-pregnane-3α,17α-diol-20-one as a key intermediate. Endocrinology 2003;144:575–580.

20 Jez JM, Penning TM: The aldo-keto reductase (AKR) superfamily: an update. Chem Biol Interact 2001;130–132:499–525.

21 Eckstein B, Borut A, Cohen S: Metabolic pathways for androstanediol formation in immature rat testis microsomes. Biochim Biophys Acta 1987;924:1–6.

22 Chase DJ, Payne AH: Changes in Leydig cell function during sexual maturation in the mouse. Biol Reprod 1983;29:1194–1200.

23 Sheffield JW, O'Shaughnessy PJ: Testicular steroid metabolism during development in the normal and hypogonadal mouse. J Endocrinol 1988;119:257–264.

24 Ge RS, Hardy MP: Variation in the end products of androgen biosynthesis and metabolism during postnatal differentiation of rat Leydig cells. Endocrinology 1998;139:3787–3795.

25 Ge RS, Hardy DO, Catterall JF, Hardy MP: Opposing changes in 3α-hydroxysteroid dehydrogenase oxidative and reductive activities in rat leydig cells during pubertal development. Biol Reprod 1999;60: 855–860.

26 Frungieri MB, Gonzalez-Calvar SI, Bartke A, Calandra RS: Influence of age and photoperiod on steroidogenic function of the testis in the golden hamster. Int J Androl 1999;22:243–252.

27 Carmichael R, Belanger A, Cusan L, Seguin C, Caron S, Labrie F: Increased testicular 5α-androstane-3α,17β-diol formation induced by treatment with [D-Ser (TBU) 6, des-Gly-NH2(10)] LHRH ethylamide in the rat. Steroids 1980;36:383–391.
28 Mahendroo MS, Wilson JD, Richardson JA, Auchus RJ: Steroid 5α-reductase 1 promotes 5α-androstane-3α,17β-diol synthesis in immature mouse testes by two pathways. Mol Cell Endocrinol 2004;222: 113–120.
29 Wilson JD, Griffin JE, Russell DW: Steroid 5α-reductase 2 deficiency. Endocr Rev 1994;14:577–593.
30 Siiteri PK, Wilson JD: Testosterone formation and metabolism during male sexual differentiation in the human embryo. J Clin Endocrinol Metab 1974; 38:113–125.
31 Gupta MK, Guryev OL, Auchus RJ: 5α-reduced C_{21} steroids are substrates for human cytochrome P450c17. Arch Biochem Biophys 2003;418:151–160.
32 Katagiri M, Kagawa N, Waterman MR: The role of cytochrome b_5 in the biosynthesis of androgens by human P450c17. Arch Biochem Biophys 1995;317: 343–347.
33 Lee-Robichaud P, Wright JN, Akhtar ME, Akhtar M: Modulation of the activity of human 17α-hydroxylase-17,20-lyase (CYP17) by cytochrome b_5:endocrinological and mechanistic implications. Biochem J 1995;308:901–908.
34 Auchus RJ: The backdoor pathway to dihydrotestosterone. Trends Endocrinol Metab 2004;15:432–438.
35 Backstrom T, Andersson A, Baird DT, Selstam G: The human corpus luteum secretes 5α-pregnane-3,20-dione. Acta Endocrinol (Copenh) 1986;111:116–121.
36 Milewich L, Mendonca BB, Arnhold I, Wallace AM, Donaldson MD, Wilson JD, Russell DW: Women with steroid 5α-reductase 2 deficiency have normal concentrations of plasma 5α-dihydroprogesterone during the luteal phase. J Clin Endocrinol Metab 1995;80:3136–3139.
37 Yokoi H, Tsuruo Y, Miyamoto T, Ishimura K: Steroid 5α-reductase type 1 immunolocalized in the adrenal gland of normal, gonadectomized, and sex hormone-supplemented rats. Histochem Cell Biol 1998;109:127–134.
38 Hanley NA, Arlt W: The human fetal adrenal cortex and the window of sexual differentiation. Trends Endocrinol Metab 2006;17:391–397.
39 Peterson RE, Imperato-McGinley J, Gautier T, Shackleton C: Male pseudohermaphroditism due to multiple defects in steroid-biosynthetic microsomal mixed-function oxidases. A new variant of congenital adrenal hyperplasia. N Engl J Med 1985;313: 1182–1191.
40 Shackleton C, Malunowicz E: Apparent pregnene hydroxylation deficiency (APHD): seeking the parentage of an orphan metabolome. Steroids 2003; 68:707–717.
41 Miller WL: Congenital adrenal hyperplasia. N Engl J Med 1986;314:1321–1322.
42 Flück CE, Tajima T, Pandey AV, Arlt W, Okuhara K, Verge CF, Jabs EW, Mendonca BB, Fujieda K, Miller WL: Mutant P450 oxidoreductase causes disordered steroidogenesis with and without Antley-Bixler syndrome. Nat Genet 2004;36:228–230.
43 Arlt W, Walker EA, Draper N, Ivison HE, Ride JP, Hammer F, Chalder SM, Borucka-Mankiewicz M, Hauffa BP, Malunowicz EM, Stewart PM, Shackleton CH: Congenital adrenal hyperplasia caused by mutant P450 oxidoreductase and human androgen synthesis: analytical study. Lancet 2004;363:2128–2135.
44 Huang N, Pandey AV, Agrawal V, Reardon W, Lapunzina PD, Mowat D, Jabs EW, Van Vliet G, Sack J, Flück CE, Miller WL: Diversity and function of mutations in P450 oxidoreductase in patients with Antley-Bixler syndrome and disordered steroidogenesis. Am J Hum Genet 2005;76:729–749.
45 Yanagibashi K, Hall PF: Role of electron transport in the regulation of the lyase activity of C-21 side-chain cleavage P450 from porcine adrenal and testicular microsomes. J Biol Chem 1986;261:8429–8433.
46 Geller DH, Auchus RJ, Mendonça BB, Miller WL: The genetic and functional basis of isolated 17,20 lyase deficiency. Nat Genet 1997;17:201–205.
47 Auchus RJ, Miller WL: Molecular modeling of human P450c17 (17α-hydroxylase/17,20-lyase): Insights into reaction mechanisms and effects of mutations. Mol Endocrinol 1999;13:1169–1182.
48 Shackleton C, Marcos J, Arlt W, Hauffa BP: Prenatal diagnosis of P450 oxidoreductase deficiency (ORD): A disorder causing low pregnancy estriol, maternal and fetal virilization, and the Antley-Bixler syndrome phenotype. Am J Med Genet A 2004;129: 105–112.
49 Homma K, Hasegawa T, Nagai T, Adachi M, Horikawa R, Fujiwara I, Tajima T, Takeda R, Fukami M, Ogata T: Urine steroid hormone profile analysis in cytochrome P450 oxidoreductase deficiency: implication for the backdoor pathway to dihydrotestosterone. J Clin Endocrinol Metab 2006;91:2643–2649.
50 Whorwood CB, Ueshiba H, del Blazo P: Plasma levels of C19 steroid glucuronides in pre-menopausal women with non-classical congenital adrenal hyperplasia. J Steroid Biochem Mol Biol 1992;42:211–221.
51 Dunaif A: Insulin resistance and the polycystic ovary syndrome: mechanism and implications for pathogenesis. Endocr Rev 1997;18:774–800.

52 Arlt W, Justl H-G, Callies F, Reincke M, Hubler D, Oettel M, Ernst M, Schulte HM, Allolio B: Oral dehydroepiandrosterone for adrenal androgen replacement: pharmacokinetics and peripheral conversion to androgens and estrogens in young healthy females after dexamethasone suppression. J Clin Endocrinol Metab 1998;83:1928–1934.
53 Ehrmann DA, Rosenfield RL, Barnes RB, Brigell DF, Sheikh Z: Detection of functional ovarian hyperandrogenism in women with androgen excess. N Engl J Med 1992;327:157–162.
54 Rosenfield RL, Barnes RB, Ehrmann DA: Studies of the nature of 17-hydroxyprogesterone hyperresonsiveness to gonadotropin-releasing hormone agonist challenge in functional ovarian hyperandrogenism. J Clin Endocrinol Metab 1994;79:1686–1692.
55 Nestler JE, Jakubowicz DJ: Decreases in ovarian cytochrome P450c17α activity and serum free testosterone after reduction of insulin secretion in polycystic ovary syndrome. N Engl J Med 1996;335:617–623.
56 Thompson DL, Horton N, Rittmaster RS: Androsterone glucuronide is a marker of adrenal hyperandrogenism in hirsute women. Clin Endocrinol (Oxf) 1990;32:283–292.
57 Falsetti L, Rosina B, De Fusco D: Serum levels of 3α-androstanediol glucuronide in hirsute and non hirsute women. Eur J Endocrinol 1998;138:421–424.
58 Fassnacht M, Schlenz N, Schneider SB, Wudy SA, Allolio B, Arlt W: Beyond adrenal and ovarian androgen generation: Increased peripheral 5α-reductase activity in women with polycystic ovary syndrome. J Clin Endocrinol Metab 2003;88:2760–2766.

Dr. Richard J. Auchus, MD, PhD
Division of Endocrinology and Metabolism
Department of Internal Medicine, University of Texas Southwestern Medical Center
5323 Harry Hines Blvd., Dallas, TX 75390-8857 (USA)
Tel. +1 214 648 6751, Fax +1 214 648 8917, E-Mail richard.auchus@UTSouthwestern.edu

Flück CE, Miller WL (eds): Disorders of the Human Adrenal Cortex.
Endocr Dev. Basel, Karger, 2008, vol 13, pp 67–81

P450 Oxidoreductase Deficiency – A New Form of Congenital Adrenal Hyperplasia

Christa E. Flück[a] · Amit V. Pandey[a] · Ningwu Huang[b] · Vishal Agrawal[b] · Walter L. Miller[b]

[a]Division of Pediatric Endocrinology and Diabetology, Department of Pediatrics, University of Bern, Bern, Switzerland; [b]Department of Pediatrics, University of California San Francisco, San Francisco, Calif., USA

Abstract

Patients with adrenal insufficiency, genital anomalies and bony malformations resembling the Antley-Bixler syndrome (a craniosynostosis syndrome), are likely to have P450 oxidoreductase (POR) deficiency. Since our first report in 2004, about 26 recessive POR mutations have been identified in 50 patients. POR is the obligate electron donor to all microsomal (type II) P450 enzymes, including the steroidogenic enzymes CYP17A1, CYP21A2 and CYP19A1. POR deficiency may cause disordered sexual development manifested as genital undervirilization in 46,XY newborns as well as overvirilization in those who are 46,XX. This may be explained by impaired aromatization of fetal androgens which may also lead to maternal virilization and low urinary estriol levels during pregnancy. A role for the alternate 'backdoor' pathway of androgen biosynthesis, leading to dihydrotestosterone production bypassing androstenedione and testosterone, has been suggested in POR deficiency but remains unclear. POR variants may play an important role in drug metabolism, as most drugs are metabolized by hepatic microsomal P450 enzymes. However, functional assays studying the effects of specific POR mutations on steroidogenesis showed that several POR variants impaired CYP17A1, CYP21A2 and CYP19A1 activities to different degrees, indicating that each POR variant must be studied separately for each potential target P450 enzyme. Thus, the impact of POR mutations on drug metabolism by hepatic P450s requires further investigation.

Human P450 Oxidoreductase Mutations: From Phenotype to Genotype

In 1985, a 46,XY patient was reported with genital ambiguity and an abnormal urinary steroid profile suggesting combined partial deficiencies of two steroidogenic enzymes CYP17A1 (17α-hydroxylase/17,20 lyase) and CYP21A2 (21-hydroxylase) [1]. Subsequently, several similar patients were described with this apparent combination of defects in enzymatic activities of CYP17A1 and CYP21A2 [2–5], but genetic analysis of the CYP17A1 and CYP21A2 genes encoding these enzymes revealed no mutations in these patients [5, 6].

Most disorders of steroidogenesis are caused by mutations in genes encoding steroidogenic enzymes, resulting in diminished or absent enzymatic activity. The only exception is congenital lipoid adrenal hyperplasia, which is caused by mutations in the steroidogenic acute regulatory protein which transports cholesterol into mitochondria to initiate steroidogenesis [7]. To explain the complex pattern of partial combined CYP17A1 and CYP21A2 deficiencies, mutations in P450 oxidoreductase (POR), the flavoprotein that supplies electrons to all microsomal P450 enzymes, were suggested [8]. However, deletion of the POR gene is embryonically lethal in mice [9, 10], which seemed to exclude POR as a candidate gene for a human disease.

Nevertheless, we found POR mutations in 4 patients who seemed to have steroid abnormalities suggesting combined defects in CYP17A1 and CYP21A2 [11]. Three of these patients had ambiguous genitalia and the skeletal malformations known as the Antley-Bixler syndrome (ABS), and one patient was an adult with steroid abnormalities, primary amenorrhea and polycystic ovaries [11]. The functional impact of the POR mutations identified in those patients was demonstrated using cytochrome c assays as well as CYP17A1 activity studies. Subsequent studies confirmed the presence of POR mutations in patients with similar patterns of steroid abnormalities with and without ABS [11–18]. To date, approximately 50 patients harboring about 26 recessive POR mutations have been described in the literature including missense and frameshift mutations, indels or splice variations (table 1). Missense mutations predominate: the mutation A287P is the most common mutation among patients of European descent and R457H is most common among patients of Japanese heritage [14]. Interestingly, 12% of the patients reported have a mutation identified on only one allele, but these patients are phenotypically indistinguishable from those with mutations on both alleles [11, 14, 15, 18]. These patients may have cryptic, unidentified mutations or they may be true manifesting heterozygotes [18].

A recent study of 842 healthy, unrelated individuals in four ethnic groups (218 African Americans, 260 Caucasian Americans, 178 Chinese Americans, and 185 Mexican Americans) identified a large number of POR sequence variations [19]. These included 140 single nucleotide polymorphisms, 13 novel missense variations, and 8 indels. In vitro analysis of the 13 novel missense mutations that were expressed in *Escherichia coli* and assayed for their abilities to support cytochrome c reduction, nicotinamide adenine dinucleotide phosphate (NADPH) oxidation, and the 17α-hydroxylase/17,20 lyase activities of CYP17A1 showed variable impairment of catalytic activities for several mutants, depending on the interacting partner. One missense mutant, A503V, was found on 27.9% of all alleles [19].

Table 1. Reported patients with POR deficiency

Patients	Chromosomal sex 46, XX/46, XY	POR mutations	Phenotype			Reference
			ABS features	abnormal genitalia	abnormal steroids	
4	2/2	7/8[1]	3	3	4	11
3	2/1	6/6	1	2	3	12
2	1/1	4/4	2	1	2	13
19 (32)[2]	6[3]/10[3]	34/38[1]	19 (32)	12[3]	10[3]	14
10	6/4	19/20[1]	9	9	10	15
3	2/1	6/6	0	2	3	16
7	2/5	14/14	5	2	7	17
1	1/0	1/2[1]	1	1	1	18
49	22/24[3]	91/98	40/49	32/49	40/49	

[1]POR mutations not identified on all alleles.
[2]Nineteen out of 32 patients with the ABS phenotype had POR mutations.
[3]Karyotype, description of genitalia or steroid profile not known for all patients.

The Antley-Bixler Syndrome

To date, about 2/3 of patients with POR deficiency were first diagnosed with ABS by geneticists. ABS is a skeletal malformation syndrome, first described in 1975 [20], characterized by craniosynostosis, midface hypoplasia, choanal atresia, radiohumeral or radioulnar synostosis, joint contractures, arachnodactyly, and bowing of the femora [21–23]. Many patients with ABS have activating mutations in the gene for fibroblast growth factor receptor 2 (FGFR2) [6, 24–27]. Fibroblast growth factors (FGFs) are mitogens involved in bone growth and development [28]. They bind to four different tyrosine kinase receptors on the cell surface (FGFRs). Dominant gain-of-function mutations in FGFRs (predominantly FGFR2) cause a variety of craniosynostosis syndromes including Pfeiffer, Apert, Jackson-Weiss, and Crouzon syndromes [29–32]. The same FGFR2 mutations can cause different clinical syndromes, suggesting that these syndromes are phenotypic variants of a single genetic disorder.

In contrast to all other craniosynostosis syndromes, about half of ABS patients have genital anomalies [6, 23]; and ABS patients with genital anomalies do not have FGFR2 mutations [6]. In a comprehensive study of 32 ABS patients with and without hormonal findings indicating a disorder of steroidogenesis, 19 patients had recessive POR mutations, 10 had dominant FGFR2 or FGFR3 mutations but no POR mutations, and no mutations in POR or FGFR were found in three [14]. This study proved that POR and FGFR2 mutations segregate completely, and that patients with an ABS-like phenotype and anomalies in genital development and steroidogenesis have a distinct disease: POR deficiency [33].

P450 Oxidoreductase and the Biochemistry of Electron Transfer

POR is an 82-kDa, membrane-associated protein first isolated as one of the components of microsomal P450 enzyme systems in 1969 [34]; the human cDNA for POR was cloned in 1989 [35], but the POR gene was characterized and sequenced only as part of the human genome project. POR is located on chromosome 7q11.23 and consists of 15 protein-coding exons spanning 32 kb, and a noncoding exon located 38.8 kb upstream [18]. The human POR gene encodes 680 amino acids, whereas the POR from rodents is 94% identical, lacks 3 amino acids at the amino terminus, and contains 677 amino acids. The N-terminus of POR has 25–30 amino-acid-long hydrophobic sequence that serves as the membrane anchor for POR to position it toward the cytoplasmic side of the endoplasmic reticulum; this sequence also plays a role in the interaction with cytochrome P450s. Deletion of this N-terminal leads to loss of POR activity. The X-ray crystal structure of a soluble form of rat POR lacking the membrane-anchoring region has been determined [36]. POR has two structurally distinct domains, an extended domain that binds reduced NADPH and flavin adenine dinucleotide (FAD) and a domain that binds flavin mononucleotide (FMN); these two domains are separated by a flexible hinge region (fig. 1).

The major function of POR is to transfer electrons from NADPH to microsomal (type II) P450 enzymes. Once the FAD molecule accepts a pair of electrons from NADPH, the hinge flexes to bring the FAD closer to the FMN domain, so that electrons can pass from FAD to FMN. The FMN domain interacts with the redox-partner binding sites of the cytochrome P450 and other electron acceptors to transfer electrons received from the FAD. Thus, unlike mitochondrial ferredoxin reductase, which only has an FAD domain to accept electrons from NADPH and needs ferredoxin as an intermediate to support type I P450 enzymes, POR is able to donate electrons directly to type II P450 enzymes [for review see 37]. The hinge region is unique to POR and provides structural flexibility to the FMN and FAD regions during electron transfer, both from NADPH and to the P450. This structural flexibility permits POR to provide electrons to a wide variety of substrates. For some P450 enzymes, cytochrome b5 may donate the second electron in the POR catalytic cycle [38], but in the case of CYP17A1, cytochrome b5 acts as an allosteric factor to improve the interaction of POR with the P450 [39]. A role for cytochrome b5 with CYP21A2 or CYP19A1 has not been described.

The catalytic activities of all microsomal P450 enzymes are POR dependent, as POR is the only protein known that can transfer electrons from NADPH to P450s.

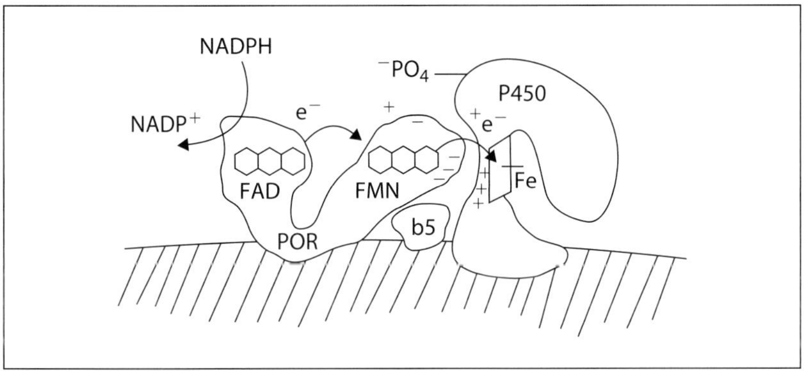

Fig. 1. Electron transfer by microsomal (type II) P450 enzymes such as CYP17A1, CYP21A2 and CYP19A1. NADPH interacts with POR, which is bound to the endoplasmic reticulum, and gives up a pair of electrons, which are received by the FAD moiety. Electron receipt elicits a conformational change, permitting the isoalloxazine rings of the FAD and FMN moieties to move closer together, so that the electrons pass from the FAD to the FMN. Following another conformational change that returns the POR to its original orientation, the FMN domain interacts with the redox-partner binding site of the P450, so that electrons reach the P450 heme iron to achieve catalysis. The interaction of POR and the P450 is coordinated by acidic residues on the surface of the FMN domain of POR, and by basic residues in the redox-partner binding site of the P450. In the case of human CYP17A1, this interaction is facilitated by the allosteric action of cytochrome b5 and by serine phosphorylation of CYP17A1, which optimize the interaction of the two proteins. Reproduced with permission from W.L. Miller.

The electron transfer reaction of POR is faster and more efficient than the P450 reactions, so that in liver and most steroidogenic tissues, the molar ratio of POR to all the microsomal P450s may be as low as 1:20; although testicular microsomes contain more POR than CYP17A1 [40]. CYP17A1 catalyzes both the 17α-hydroxylation required for the production of 17-hydroxy, 21-carbon precursors of cortisol, and the 17,20 lyase activity needed for the synthesis of C19 precursors of sex steroids, but an increase in the ratio of POR to CYP17A1 enhances the 17,20 lyase reaction far more than the 17α-hydroxylase reaction [40, 41]. The affinity of interaction of POR with CYP17A1 and some hepatic drug-metabolizing P450 enzymes can be modulated by the allosteric action of cytochrome b5 [42, 43].

P450 Oxidoreductase and Cytochrome P450s

Cytochrome P450 enzymes are heme-containing proteins that catalyze a broad range of oxidative reactions (http://drnelson.utmem.edu/CytochromeP450.html). The 57 known human P450 enzymes are grouped in type I P450s, which are found in the mitochondria, and the type II P450s found in the endoplasmic reticulum. There are seven type I P450s, which receive electrons from NADPH via the coupled electron

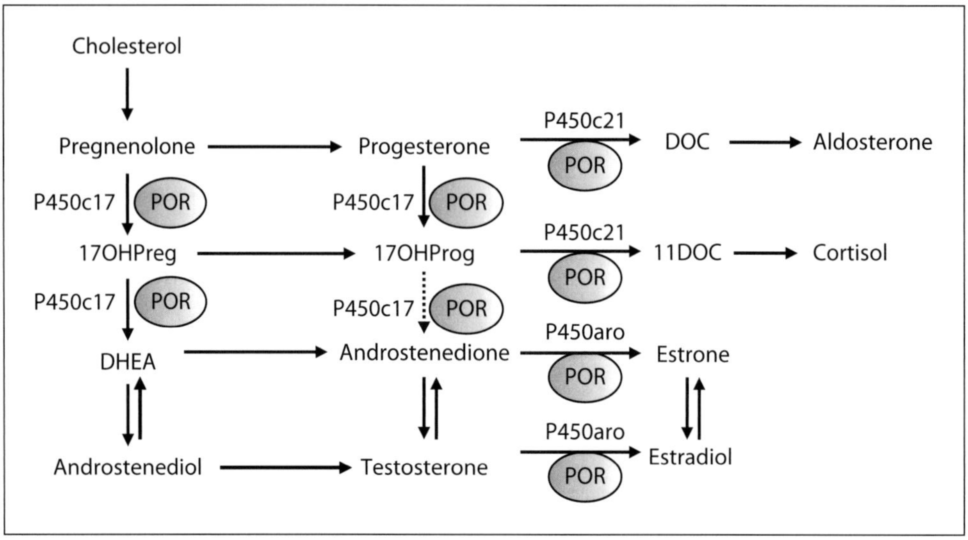

Fig. 2. Scheme of the steroid pathway. Enzymes that require POR as an electron donor are depicted; enzymes that do not require POR are simply depicted as arrows. Note that the human 17,20 lyase activity of CYP17A1 converts 17α-hydroxypregnenolone (17OHPreg) to DHEA but does not effectively convert 17α-hydroxyprogesterone (17OHProg) to androstenedione. DOC = 11-deoxycorticosterone; 11DOC = 11-deoxycortisol.

chain of ferredoxin reductase and ferredoxin; all type I P450 enzymes are involved in the synthesis of steroids, sterols and bile acids [37]. By contrast, there are 50 type II P450s, which receive electrons from NADPH via a single redox partner, POR (fig. 1) [37]. The three steroidogenic type II P450s enzymes are CYP17A1, CYP21A2 and CYP19A1 (fig. 2). Other type II P450s are involved in drug metabolism or in the biosynthetic pathways leading to cholesterol, bile acids and eicosanoids, while some remain 'orphans', having unknown activities.

Impact of P450 Oxidoreductase Deficiency on Steroidogenesis

Patients with POR deficiency have a complex pattern of disordered steroidogenesis. The steroid metabolizing reactions catalyzed by CYP17A1, CYP21A2 and CYP19A1 require POR for electron transfer (fig. 1). Loss of the 17α-hydroxylase and 17,20 lyase activities of CYP17A1 explains the elevated plasma concentrations of deoxycorticosterone and corticosterone but the decreased levels of cortisol and C19 steroids (dehydroepiandrosterone – DHEA, DHEA sulfate and androstenedione). These abnormalities may not be noticed at the basal state but are more readily seen following stimulation with adrenocorticotropic hormone. Loss of 21-hydroxylase activity results in diminished plasma cortisol (and rarely aldosterone) but elevated 21-deoxycortisol concentrations, especially after adrenocorticotropic hormone stimulation. However, in contrast to isolated 21-hydroxylase deficiency due to CYP21A2 mutations, POR patients typically have only moderately elevated plasma 17α-hydroxyprogesterone, and decreased rather than increased C19 steroid concentrations due to the combined deficiency of CYP21A2 and CYP17A1.

Impairment of CYP19A1 activity due to POR deficiency will result in diminished production of estrogens from androgen precursors in the ovaries of affected individuals. Impairment of aromatase activity can affect the conversion of fetal C19 androgen precursors (DHEA and DHEA sulfate) to estriol and estrone in the fetoplacental unit, causing virilization of the fetus and the mother (manifested by acne, voice changes, and hirsutism). Although 46,XX fetuses with aromatase (CYP19A1) gene mutations typically develop severely virilized external genitalia during pregnancy, the affected 46,XY patients do not have genital anomalies [44]. So there is an inconsistency in the undervirilization of some males with POR deficiency and the mild virilization of some females with POR deficiency. This apparent paradox may be explained by the 'backdoor pathway' of androgen biosynthesis, which is the main pathway to dihydrotestosterone production in tammar wallaby pouch young [45]. In this pathway, dihydrotestosterone is formed from 17α-hydroxyprogesterone via 5α reductase, bypassing the conventional intermediates androstenedione and testosterone [45]. The role of this pathway in human androgen production remains unclear. From the urinary steroid analyses there is evidence both for [17, 46] and against [17, 47] its activity in patients with POR deficiency. The 'backdoor pathway' is summarized in the chapter by Ghayee and Auchus [pp. 55–66].

The activities of most POR mutations have been tested by assays based on CYP17A1 and cytochrome c [11, 14, 19], and for some with all three microsomal P450s involved in steroidogenesis, using recombinant proteins produced in yeast or bacteria [11, 12, 14, 19, 48, 49]. In our first report, we tested the impact of POR variants A287P, R457H, V492E, C569Y and V608F on both 17α-hydroxylase and 17,20 lyase activities of CYP17A1 and compared the results with the cytochrome c reduction/NADPH oxidation assays [11]. We found a good correlation between the patients' clinical features and the functional assays based on CYP17A1, but a poorer correlation with the assays based on cytochrome c. We made three important observations. First, mutations in the FAD region that destroy the binding of FAD to POR resulted in almost total loss of activity in all assays. Second, the A287P mutation, which is not directly involved in the electron transport but located close to the hinge region (residues 234–283), had reduced CYP17A1 activities but nearly wild-type (WT) activity in the cytochrome c assays. Third, mutations C569Y and V608F, which are located in the NADPH binding domain, had higher Michaelis constants for NADPH in the cytochrome c assay, suggesting that these mutants might affect binding of NADPH to POR. Subsequent studies have tested the impact of many known missense mutations on CYP17A1 activities (fig. 3) [14, 19]. So far,

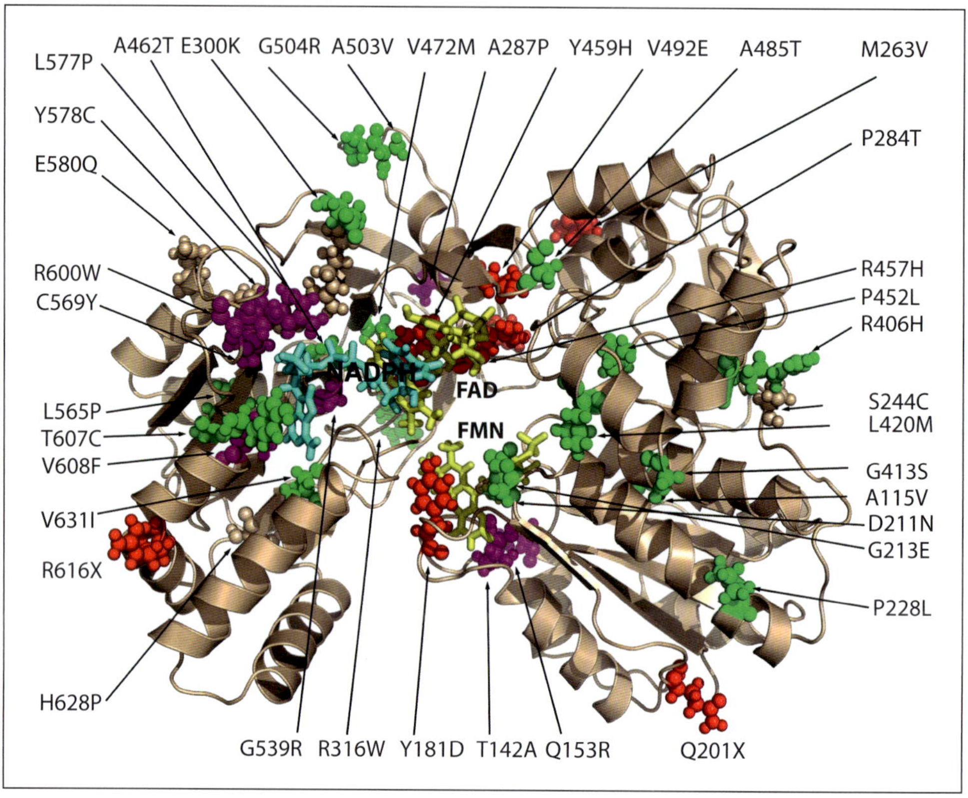

Fig. 3. Model of human POR protein, showing all identified missense mutations. The model has been described [14] and is based on the crystal structure of rat POR lacking 65 amino terminal residues [36]. The α-carbon backbone is depicted as a narrow ribbon. Ball-and-stick models are used to represent the FAD and FMN moieties (yellow), and the NADPH (cyan). The missense residues identified by sequencing are depicted by charged packed sphere images of different colors, corresponding to their V_{max}/K_m for CYP17A1 activities: red: $<$25%; magenta: 25–50%; green: $>$50%.

we have studied the enzymology of POR variants P55L, A115V, T142A, Q153R, Y181D, D211N, G213E, P228L, M263V, P284L, P284T, A287P, E300K, R316W, R406H, G413S, P452L, R457H, Y459H, A462T, V472M, A485T, V492E, A503V, G504R, G539R, L565P, C569Y, R600W, Y607C, V608F, R616X, V631I, and F646del [14, 19].

We have also compared the effects of POR variants on CYP19A1 activity with the effects on CYP17A1 activities [48]. The mutations R457H and V492E caused a complete loss of CYP19A1 activity, confirming that POR mutations disrupting electron transfer will severely affect any interacting P450. In contrast, POR mutants A287P, C569Y and V608F, had variable effects on CYP19A1 and CYP17A1 activities (fig. 4). POR mutations C569Y and V608F, which were identified in a compound heterozygote

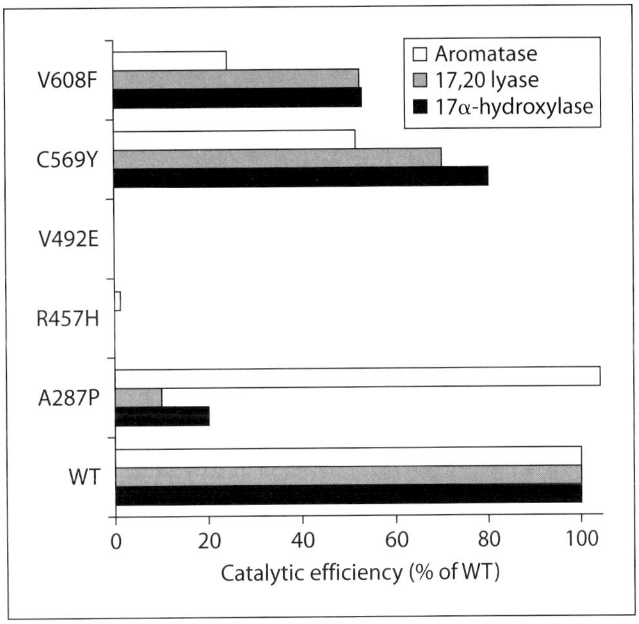

Fig. 4. Differential impact of POR missense mutations on CYP17A1 versus CYP19A1 activities. Both CYP17A1 and CYP19A1 activities were assessed in vitro when supported by either WT or mutant POR proteins. Proteins for the assays were produced in a humanized yeast system [11, 48]. Catalytic efficiencies were calculated as V_{max}/K_m and are given as percentage of WT.

state in a mildly affected female patient with a polycystic ovary syndrome-like phenotype, had more than 50% of WT CYP17A1 activities but less than 50% of WT CYP19A1 activity. In addition, those mutants seemed to be more sensitive to the amount of NADPH available. Conversely, the A287P mutant did not affect CYP19A1 activity but reduced CYP17A1 activities remarkably.

Molecular modeling and protein docking studies suggested that A287P disrupts the interaction of POR with CYP17A1 but not the interaction of POR with CYP19A1. We also determined whether pH and salt concentrations may affect POR-P450 reaction kinetics and studied the effects of pH and KCl on mutant and WT POR interacting with CYP19A1. Octanol was used to study the effects of disordered or rearranged membrane structure on the rates of CYP19A1 activity supported by WT or mutant POR [48]. The differential inhibition of POR WT and A278P, S244C, H628P and Y181D mutants on enzymatic activity of CYP21A2 was studied by others [49]. Mutation Y181D, which is involved in the FMN binding, impaired both CYP21A2 and CYP17A1 activities similarly, but the A287P mutant impaired CYP17A1 activities but did not affect CYP21A2 activity. The basis of the selectivity of some POR variants towards certain P450s or other electron acceptors will become clearer as studies with other target proteins are carried out. The impact of specific POR variants that do not lead to obvious changes disrupting the structure, such as loss of FAD or FMN or a truncated protein, cannot be predicted from a single cytochrome c or P450 assay but need to be tested with each P450 and interaction partner specifically.

Impact of P450 Oxidoreductase on Hepatic P450s

Many type II P450 enzymes are involved in hepatic drug metabolism, and the majority of drugs in clinical use are metabolized by these enzymes [50]. Sequence variations in these P450 enzymes cause variations in drug responses both among individuals and among ethnic groups. As all type II P450s depend on POR for electron supply, POR variants may also contribute to variations in drug metabolism. Before POR mutations were identified in patients with ABS and disordered steroidogenesis, patients with a similar phenotype were reported after in utero exposure to fluconazole, suggesting that this drug might be teratogenic [51–53]. Fluconazole is an inhibitor of lanosterol 14α-demethylase (CYP51) which converts lanosterol to ergosterol in the cholesterol biosynthetic pathway; a genetic defect in CYP51 was postulated but not found [54]. Nevertheless, lymphoblast cells isolated from POR patients have higher lanosterol levels than cells from controls or from patients with ABS due to FGFR2 mutations [54], probably because POR mutations affect CYP51 enzyme activity. In fact, one of these patients exposed to fluconazole in utero [54] was later found to harbor homozygote POR mutations A287P [11]. Fluconazole also inhibits the hepatic drug metabolizing type II P450s CYP2C9, CYP2C19 and CYP3A4 [55], which may be affected by POR variants.

Further evidence for the importance of POR in hepatic P450 activities comes from a liver-specific knockout of POR in mice [56]. Homozygous deletion of the POR gene (called Cpr in mice) in the liver decreased microsomal cytochrome P450 and heme oxygenase activities, and decreased pentobarbital clearance and total plasma cholesterol. However, unlike global POR knockout mice, which are embryonic lethal [9], liver-specific knockouts are viable, and are normal in gross appearance, growth, and fertility [56].

Some functional data concerning the impact of POR variants on hepatic drug-metabolizing P450s are now emerging. The activities of 10 P450 enzymes (CYP1A2, 2A6, 2B6, 2C8, 2C9, 2C19, 2D6, 2E1, 3A4/5 and 4A9/11) were examined in pooled liver microsome samples using test drugs that are specific for each enzyme [57]. However, since P450 genetic polymorphisms are quite common, some of the variations reported may be due to variations in the P450 rather than in POR. Also, as the liver samples were collected at different times and were frozen from 1 to 36 h, loss of POR and P450 activities might reflect variations in handling the samples. The 35 POR mutants identified from patients and from the sequencing of 842 POR genes from normal individuals [19] have now been expressed in bacteria and their activities measured in vitro with bacterially expressed CYP1A2 and CYP2C19, again showing that some POR variants will affect the activity of some P450 enzymes, but not others [58].

Skeletal Malformations in P450 Oxidoreductase Deficiency

More than 2/3 of patients with POR mutations have skeletal malformations that are indistinguishable from those in ABS caused by FGFR2 mutations. However, the pathogenesis underlying these bone malformations is unknown. Several lines of evidence suggest a possible role for cholesterol biosynthesis and POR in bone development. First, cholesterol biosynthesis requires squalene epoxidase, a non-P450 enzyme, and 14α-demethylase (CYP51), both of which require POR for electron transfer [59, 60]. Second, human disorders of cholesterol biosynthesis can cause skeletal anomalies; an example is the Smith-Lemli-Opitz syndrome, which is caused by mutations in the gene for 7-dehydrocholesterol reductase. Third, cholesterol is required for normal activity and signal transduction by hedgehog proteins, which are crucial for the regulation of growth, morphogenesis and bone formation in embryogenesis [61]. In addition, retinoic acid might be involved in the development of skeletal malformations in POR deficiency. Inappropriate levels (excess or deficiency) of retinoic acid can cause a wide range of limb malformations [62]. Retinoic acid is metabolized by microsomal CYP26A1, which depends on POR for electron transfer. Reducing retinoic acid in the diet of pregnant mice carrying POR knockout pups partially ameliorated their phenotype [10], suggesting a role for retinoic acid toxicity in bone malformations of POR deficiency. However, to date all of these suggestions remain hypotheses that need to be addressed in future studies.

Conclusions and Perspectives

The discovery of POR deficiency has broadened the spectrum of congenital adrenal hyperplasia. The typical POR-deficient patient presents with genital and skeletal anomalies consistent with the ABS phenotype, and has a complex pattern of abnormalities in the steroid hormone profile suggesting impaired CYP17A1, CYP21A2 and CYP19A1 activities. However, the clinical spectrum of POR deficiency is quite broad, and ranges from severely handicapped ABS individuals to mildly affected normal-looking adults with compromised fertility. Thus it is difficult to estimate the prevalence of POR deficiency. Whether or not POR deficiency is common will only become apparent after the functional activities of POR variants found in normal population are characterized with different enzymatic targets. The diagnosis of POR deficiency may be considered from clinical and hormonal characteristics, but requires confirmation by genetic analysis. Treatment may include replacement of glucocorticoids, sex steroids and mineralocorticoids as assessed by low basal or stimulated serum hormone levels for each individual patient. The skeletal malformations of POR deficiency require orthopedic management, and most reported mortality is related to bony abnormalities causing respiratory problems (e.g. choanal obstruction). Another potentially important clinical issue remains the effect of POR variants on drug-metabolizing P450 enzymes. As a first step towards pharmacogenetic testing for relevant POR variants, detailed analysis of the POR gene in 842 normal subjects showed that there is a common POR variant, A503V, found in about 28% of alleles [19]. Testing of 35 missense mutations by both cytochrome c assays and CYP17A1 assays revealed that several of these mutations had variations in activities depending on substrates, indicating that each POR mutant must be assayed separately with potential target P450. These 35 POR sequence variants identified have now be tested with two drug-metabolizing P450 enzymes [58], and further data should be available soon.

Acknowledgements

This work was supported by grants from the Swiss National Science Foundation (320000-116299 to C.E.F. and 3100A0-113719 to A.V.P.) and from National Institutes of Health grants HD41959 and GM73020 to W.L.M.

References

1 Peterson RE, Imperato-McGinley J, Gautier T, Shackleton CHL: Male pseudohermaphroditism due to multiple defects in steroid-biosynthetic microsomal mixed-function oxidases. A new variant of congenital adrenal hyperplasia. N Engl J Med 1985;313:1182–1191.

2 Malunowicz E, Romer TE, Szarras-Czapnik M, Mielniczuk Z, Gajewska D: Combined deficiency of 17α-hydroxylase and 21-hydroxylase in an 8 years old girl. Endokrynol Pol 1987;38:117–124.

3 Augarten A, Pariente C, Gazit E, Chayen R, Goldfarb H, Sack J: Ambiguous genitalia due to partial activity of cytochromes P450c17 and P450c21. J Steroid Biochem Mol Biol 1992;41:37–41.

4 Leiberman E, Hershkovitz E, Lauber-Biason A, Phillip M, Zachmann M: Subnormal cortisol response to adrenocorticotropin in isolated partial 17,20-lyase deficiency. J Pediatr Endocrinol Metab 1997;10:387–390.

5 Adachi M, Tachibana K, Asakura Y, Suwa S, Nishimura G: A male patient presenting with major clinical symptoms of glucocorticoid deficiency and skeletal dysplasia, showing a steroid pattern compatible with 17α-hydroxylase/17,20-lyase deficiency, but without obvious CYP17 gene mutations. Endocr J 1999;46:285–292.

6 Reardon W, Smith A, Honour JW, Hindmarsh P, Das D, Rumsby G, Nelson I, Malcolm S, Ades L, Sillence D, Kumar D, DeLozier-Blanchet C, McKee S, Kelly T, McKeehan WL, Baraitser M, Winter RM: Evidence for digenic inheritance in some cases of Antley-Bixler syndrome? J Med Genet 2000;37:26–32.

7 Miller WL: Congenital lipoid adrenal hyperplasia: the human gene knockout for the steroidogenic acute regulatory protein. J Mol Endocrinol 1997;19:227–240.

8 Miller WL: Congenital adrenal hyperplasia. N Engl J Med 1986;314:1321–1322.

9 Shen AL, O'Leary KA, Kasper CB: Association of multiple developmental defects and embryonic lethality with loss of microsomal NADPH-cytochrome P450 oxidoreductase. J Biol Chem 2002;277:6536–6541.

10 Otto DM, Henderson CJ, Carrie D, Davey M, Gundersen TE, Blomhoff R, Adams RH, Tickle C, Wolf CR: Identification of novel roles of the cytochrome P450 system in early embryogenesis: effects on vasculogenesis and retinoic acid homeostasis. Mol Cell Biol 2003;23:6103–6116.

11 Flück CE, Tajima T, Pandey AV, Arlt W, Okuhara K, Verge CF, Jabs EW, Mendonca BB, Fujieda K, Miller WL: Mutant P450 oxidoreductase causes disordered steroidogenesis with and without Antley-Bixler syndrome. Nat Genet 2004;36:228–230.

12 Arlt W, Walker EA, Draper N, Ivison HE, Ride JP, Hammer F, Chalder SM, Borucka-Mankiewicz M, Hauffa BP, Malunowicz EM, Stewart PM, Shackleton CH: Congenital adrenal hyperplasia caused by mutant P450 oxidoreductase and human androgen synthesis: analytical study. Lancet 2004;363:2128–2135.

13 Adachi M, Tachibana K, Asakura Y, Yamamoto T, Hanaki K, Oka A: Compound heterozygous mutations of cytochrome P450 oxidoreductase gene (POR) in two patients with Antley-Bixler syndrome. Am J Med Genet 2004;128A:333–339.

14 Huang N, Pandey AV, Agrawal V, Reardon W, Lapunzina PD, Mowat D, Jabs EW, Van Vliet G, Sack J, Flück CE, Miller WL: Diversity and function of mutations in P450 oxidoreductase in patients with Antley-Bixler syndrome and disordered steroidogenesis. Am J Hum Genet 2005;76:729–749.

15 Fukami M, Horikawa R, Nagai T, Tanaka T, Naiki Y, Sato N, Okuyama T, Nakai H, Soneda S, Tachibana K, Matsuo N, Sato S, Homma K, Nishimura G, Hasegawa T, Ogata T: Cytochrome P450 oxidoreductase gene mutations and Antley-Bixler syndrome with abnormal genitalia and/or impaired steroidogenesis: molecular and clinical studies in 10 patients. J Clin Endocrinol Metab 2005;90:414–426.

16 Fukami M, Hasegawa T, Horikawa R, Ohashi T, Nishimura G, Homma K, Ogata T: Cytochrome P450 oxidoreductase deficiency in three patients initially regarded as having 21-hydroxylase deficiency and/or aromatase deficiency: diagnostic value of urine steroid hormone analysis. Pediatr Res 2006;59:276–280.

17 Homma K, Hasegawa T, Nagai T, Adachi M, Horikawa R, Fujiwara I, Tajima T, Takeda R, Fukami M, Ogata T: Urine steroid hormone profile analysis in cytochrome P450 oxidoreductase deficiency: implication for the backdoor pathway to dihydrotestosterone. J Clin Endocrinol Metab 2006;91:2643–2649.

18 Scott RR, Gomes LG, Huang N, Van Vliet G, Miller WL: Apparent manifesting heterozygosity in P450 oxidoreductase deficiency and its effect on coexisting 21-hydroxylase deficiency. J Clin Endocrinol Metab 2007;92:2318–2322.

19 Huang N, Agrawal V, Giacomini KM, Miller WL: Pharmacogenetics of P450 oxidoreductase. Genetic variation in 842 individuals from four ethnic groups and enzymatic activity of 15 missense mutations. Proc Natl Acad Sci USA 2008;105:1733–1738.

20 Antley R, Bixler D: Trapezoidocephaly, midfacial hypoplasia and cartilage abnormalities with multiple synostoses and skeletal fractures. Birth Defects Orig Artic Ser 1975;11:397–401.

21 DeLozier CD, Antley RM, Williams R, Green N, Heller RM, Bixler D, Engel E: The syndrome of multisynostotic osteodysgenesis with long-bone fractures. Am J Med Genet 1980;7:391–403.

22 Hassell S, Butler MG: Antley-Bixler syndrome: report of a patient and review of literature. Clin Genet 1994;46:372–376.

23 Crisponi G, Porcu C, Piu ME: Antley-Bixler syndrome: case report and review of the literature. Clin Dysmorphol 1997;6:61–68.

24 Chun K, Siegel-Bartelt J, Chitayat D, Phillips J, Ray PN: FGFR2 mutation associated with clinical manifestations consistent with Antley-Bixler syndrome. Am J Med Genet 1998;77:219–224.

25 Gripp KW, Stolle CA, McDonald-McGinn DM, Markowitz RI, Bartlett SP, Katowitz JA, Muenke M, Zackai EH: Phenotype of the fibroblast growth factor receptor 2 Ser351Cys mutation: Pfeiffer syndrome type III. Am J Med Genet 1998;78:356–360.

26 Pulleyn LJ, Reardon W, Wilkes D, Rutland P, Jones BM, Hayward R, Hall CM, Brueton L, Chun N, Lammer E, Malcolm S, Winter RM: Spectrum of craniosynostosis phenotypes associated with novel mutations at the fibroblast growth factor receptor 2 locus. Eur J Hum Genet 1996;4:283–291.

27 Schaefer F, Anderson C, Can B, Say B: Novel mutation in the FGFR2 gene at the same codon as the Crouzon syndrome mutations in a severe Pfeiffer syndrome type 2 case. Am J Med Genet 1998;75:252–255.

28 Cohen MM Jr: The new bone biology: pathologic, molecular, and clinical correlates. Am J Med Genet A 2006;140:2646–2706.

29 Muenke M, Schell U, Hehr A, Robin NH, Losken HW, Schinzel A, Pulleyn LJ, Rutland P, Reardon W, Malcolm S, Winter RM: A common mutation in the fibroblast growth factor receptor 1 gene in Pfeiffer syndrome. Nat Genet 1994;8:269–274.

30 Reardon W, Winter RM, Rutland P, Pulleyn LJ, Jones BM, Malcolm S: Mutations in the fibroblast growth factor receptor 2 gene cause Crouzon syndrome. Nat Genet 1994;8:98–103.

31 Rutland P, Pulleyn LJ, Reardon W, Baraitser M, Hayward R, Jones B, Malcolm S, Winter RM, Oldridge M, Slaney SF, et al: Identical mutations in the FGFR2 gene cause both Pfeiffer and Crouzon syndrome phenotypes. Nat Genet 1995;9:173–176.

32 Park WJ, Meyers GA, Li X, Theda C, Day D, Orlow SJ, Jones MC, Jabs EW: Novel FGFR2 mutations in Crouzon and Jackson-Weiss syndromes show allelic heterogeneity and phenotypic variability. Hum Mol Genet 1995;4:1229–1233.

33 Miller WL: P450 oxidoreductase deficiency: a new disorder of steroidogenesis with multiple clinical manifestations. Trends Endocrinol Metab 2004;15:311–315.

34 Lu AY, Junk KW, Coon MJ: Resolution of the cytochrome P-450-containing ω-hydroxylation system of liver microsomes into three components. J Biol Chem 1969;244:3714–3721.
35 Yamano S, Aoyama T, McBride OW, Hardwick JP, Gelboin HV, Gonzalez FJ: Human NADPH-P450 oxidoreductase: complementary DNA cloning, sequence and vaccinia virus-mediated expression and localization of the CYPOR gene to chromosome 7. Mol Pharmacol 1989;36:83–88.
36 Wang M, Roberts DL, Paschke R, Shea TM, Masters BSS, Kim JJ: Three-dimensional structure of NADPH-cytochrome P450 reductase: prototype for FMN- and FAD-containing enzymes. Proc Natl Acad Sci USA 1997;94:8411–8416.
37 Miller WL: Minireview: regulation of steroidogenesis by electron transfer. Endocrinology 2005;146: 2544–2550.
38 Guengerich FP, Johnson WW: Kinetics of ferric cytochrome P450 reduction by NADPH-cytochrome P450 reductase: rapid reduction in the absence of substrate and variations among cytochrome P450 systems. Biochemistry 1997;36:14741–14750.
39 Auchus RJ, Lee TC, Miller WL: Cytochrome b5 augments the 17,20-lyase activity of human P450c17 without direct electron transfer. J Biol Chem 1998; 273:3158–3165.
40 Yanagibashi K, Hall PF: Role of electron transport in the regulation of the lyase activity of C21 side-chain cleavage P-450 from porcine adrenal and testicular microsomes. J Biol Chem 1986;261:8429–8433.
41 Lin D, Black SM, Nagahama Y, Miller WL: Steroid 17α-hydroxylase and 17,20-lyase activities of P450c17: contributions of serine106 and P450 reductase. Endocrinology 1993;132:2498–2506.
42 Yamazaki H, Nakano M, Imai Y, Ueng YF, Guengerich FP, Shimada T: Roles of cytochrome b5 in the oxidation of testosterone and nifedipine by recombinant cytochrome P450 3A4 and by human liver microsomes. Arch Biochem Biophys 1996;325: 174–182.
43 Loughran PA, Roman LJ, Miller RT, Masters BS: The kinetic and spectral characterization of the E. coli-expressed mammalian CYP4A7:cytochrome b5 effects vary with substrate. Arch Biochem Biophys 2001;385:311–321.
44 Grumbach MM, Auchus RJ: Estrogen: Consequences and implications of human mutations in synthesis and action. J Clin Endocrinol Metab 1999; 84:4677–4694.
45 Auchus RJ: The backdoor pathway to dihydrotestosterone. Trends Endocrinol Metab 2004;15:432–438.
46 Shackleton C, Marcos J, Arlt W, Hauffa BP: Prenatal diagnosis of P450 oxidoreductase deficiency (ORD): a disorder causing low pregnancy estriol, maternal and fetal virilization, and the Antley-Bixler syndrome phenotype. Am J Med Genet 2004;129A: 105–112.
47 Hershkovitz E, Parvari R, Wudy SA, Hartmann MF, Gomes LG, Loewental N, WL M: Apparent isolated 17,20 lyase deficiency caused by the homozygous mutation G539R in P450 oxidoreductase, (submitted).
48 Pandey AV, Kempna P, Hofer G, Mullis PE, Fluck CE: Modulation of human CYP19A1 activity by mutant NADPH P450 oxidoreductase. Mol Endocrinol 2007;21:2579–2595.
49 Dhir V, Ivison HE, Krone N, Shackleton CHL, Doherty AJ, Stewart PM, Arlt W: Differential inhibition of CYP17A1 and CYP21A2 activities by the P450 oxidoreductase mutant A287P. Mol Endocrinol 2007;21:1958–1968.
50 Ingelman-Sundberg M: Pharmacogenetics of cytochrome P450 and its applications in drug therapy: the past, present and future. Trends Pharmacol Sci 2004;25:193–200.
51 Lee BE, Feinberg M, Abraham JJ, Murthy AR: Congenital malformations in an infant born to a woman treated with fluconazole. Pediatr Infect Dis J 1992;11:1062–1064.
52 Pursley TJ, Blomquist IK, Abraham J, Andersen HF, Bartley JA: Fluconazole-induced congenital anomalies in three infants. Clin Infect Dis 1996;22:336–340.
53 Aleck KA, Bartley DL: Multiple malformation syndrome following fluconazole use in pregnancy: report of an additional patient. Am J Med Genet 1997;72:253–256.
54 Kelley RI, Kratz LE, Glaser RL, Netzloff ML, Wolf LM, Jabs EW: Abnormal sterol metabolism in a patient with Antley-Bixler syndrome and ambiguous genitalia. Am J Med Genet 2002;110:95–102.
55 Niwa T, Shiraga T, Takagi A: Effect of antifungal drugs on cytochrome P450 (CYP) 2C9, CYP2C19, and CYP3A4 activities in human liver microsomes. Biol Pharm Bull 2005;28:1805–1808.
56 Gu J, Weng Y, Zhang QY, Cui H, Behr M, Wu L, Yang W, Zhang L, Ding X: Liver-specific deletion of the NADPH-cytochrome P450 reductase gene: impact on plasma cholesterol homeostasis and the function and regulation of microsomal cytochrome P450 and heme oxygenase. J Biol Chem 2003;278: 25895–25901.
57 Hart SN, Wang S, Nakamoto K, Wesselman C, Li Y, Zhong XB: Genetic polymorphisms in cytochrome P450 oxidoreductase influence microsomal P450-catalyzed drug metabolism. Pharmacogenet Genomics 2008;18:11–24.

58 Agrawal V, Huang N, WL M: Pharmacogenetics of P450 oxidoreductase. Effect of sequence variants on activities of CYP1A2 and CYP2C19. Pharmacogenet Genomics; in press.
59 Ono T, Bloch K: Solubilization and partial characterization of rat liver squalene epoxidase. J Biol Chem 1975;250:1571–1579.
60 Debeljak N, Fink M, Rozman D: Many facets of mammalian lanosterol 14a-demethylase from the evolutionarily conserved cytochrome P450 family CYP51. Arch Biochem Biophys 2003;409:159–171.
61 Gofflot F, Hars C, Illien F, Chevy F, Wolf C, Picard JJ, Roux C: Molecular mechanisms underlying limb anomalies associated with cholesterol deficiency during gestation: implications of Hedgehog signaling. Hum Mol Genet 2003;12:1187–1198.
62 Lee GS, Kochhar DM, Collins MD: Retinoid-induced limb malformations. Curr Pharm Des 2004;10: 2657–2699.

Dr. Christa E. Flück
Pediatric Endocrinology and Diabetology, University Children's Hospital Bern
Freiburgstrasse 15, G3 812
CH–3010 Bern (Switzerland)
Tel. +41 31 632 04 99, Fax +41 31 632 84 24, E-Mail christa.flueck@dkf.unibe.ch

Flück CE, Miller WL (eds): Disorders of the Human Adrenal Cortex.
Endocr Dev. Basel, Karger, 2008, vol 13, pp 82–98

Long-Term Outcome of Prenatal Treatment of Congenital Adrenal Hyperplasia

Svetlana Lajic · Anna Nordenström · Tatja Hirvikoski

Department of Molecular Medicine and Surgery, Karolinska Institutet, Stockholm, Sweden

Abstract

Prenatal treatment of congenital adrenal hyperplasia (CAH) with dexamethasone to minimize the genital virilization of external genitalia of affected girls has been in use since the mid-1980s. The positive effect of reducing virilization is now established. However, experimental data from animal studies and observations on adverse medical events in human newborns have raised concerns about the long-term safety of the treatment. Most animal studies on prenatal treatment with synthetic glucocorticoids have been designed to mimic treatment for lung maturation in preterm infants. The primary focus has been on a possible impact on fetal programming and the development of the metabolic syndrome with insulin resistance, type 2 diabetes, and high blood pressure. Altered reactivity to stress as a function of differences in reactivity of the HPA axis and glucocorticoid receptor function have been assayed. Effects on cognition, especially memory, have been observed. In children at risk for CAH and treated prenatally with dexamethasone, no overall effects on full-scale IQ have been observed, but a negative effect on verbal working memory has been reported. Contradictory effects on social behavior with respect to shyness and inhibition have been discussed. There is an urgent need for in-depth studies of long-term outcome in prenatal treatment of CAH regarding both maternal side effects and possible negative metabolic as well as cognitive and behavioral effects in the growing fetus and the child in her development into adulthood.

Clinical Aspects of 21-Hydroxylase Deficiency

Congenital adrenal hyperplasia (CAH) refers to a group of recessively inherited disorders of adrenal steroidogenesis. Cortisol synthesis is reduced due to defects in one of five different enzymes (fig. 1). The most common cause of CAH is either a 21-hydroxylase deficiency (21OHD; >90%, incidence 1:14,000 in most populations) [1] or an 11β-hydroxylase deficiency (<5%, incidence ~1:100,000) both of which result in virilization of severely affected female fetuses. Virilization is due to accumulation of androgen precursors (dehydroepiandrosterone – DHEA/DHEA sulfate – DHEAS, androstenedione) produced in the adrenal cortex (due to the increased adrenocorticotropic hormone, ACTH, drive) and their conversion to the potent androgens testosterone

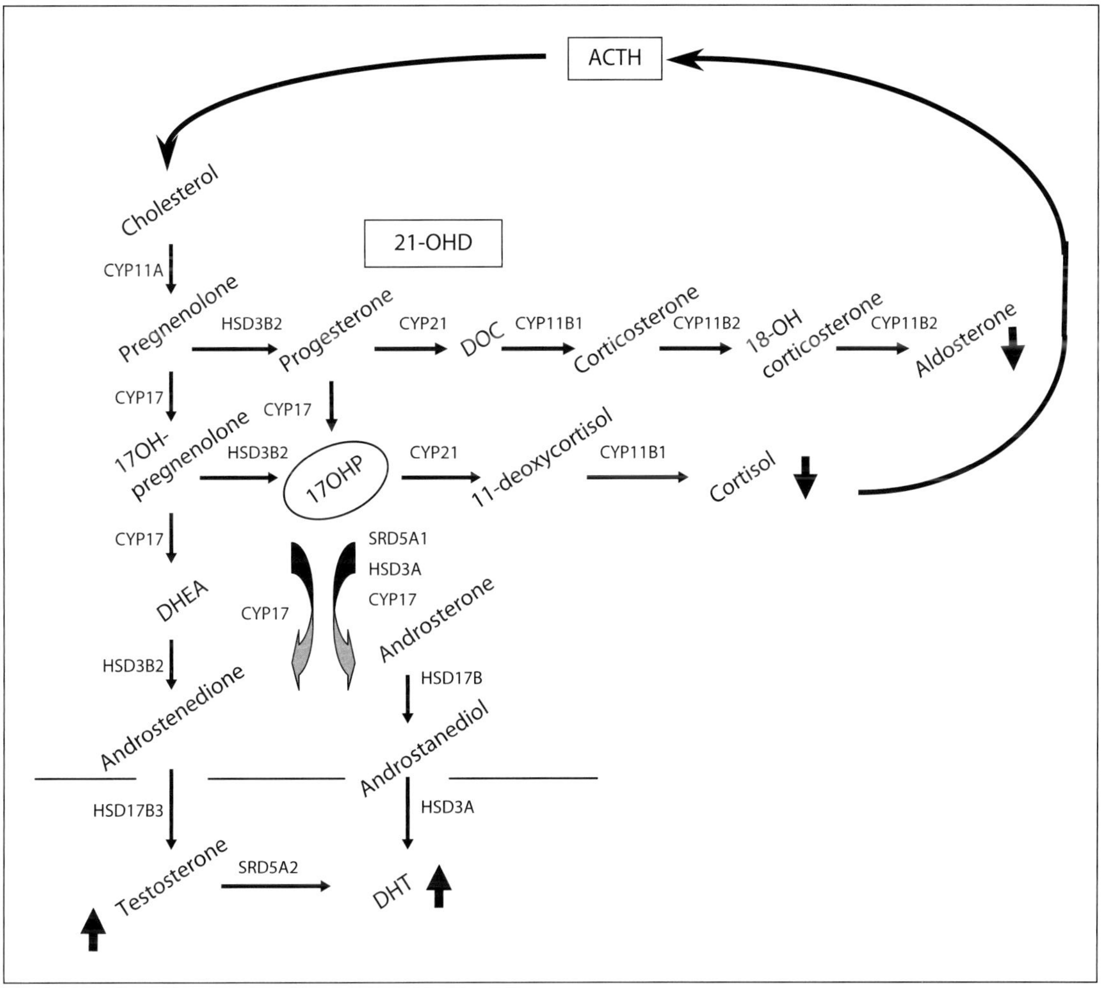

Fig. 1. Steroidogenesis in 21OHD.

and dihydrotestosterone. The recently described 'backdoor' pathway descending from 17-hydroxyprogesterone might lead to additional synthesis of androgens (androstanediol) that are converted in the external genitalia to dihydrotestosterone, thus further aggravating in utero virilization in girls with 21OHD [2, 3].

Three out of 4 patients with virilizing 21OHD have the most severe, salt-wasting (SW) form of CAH with very low or no production of aldosterone and cortisol. The newborn child presents with vomiting, lethargy and failure to thrive during the first weeks of life and unless treated with glucocorticoids (GCs) will develop life-threatening salt loss and hypotonic shock [4]. Girls are born severely virilized and are thus more likely to be diagnosed. The development of neonatal screening programs for 21OHD

Fig. 2. Common mutations giving rise to 21OHD in relation to clinical severity and enzyme activity in vitro. A group of completely inactivating mutations (Null) together with the I2 splice and I172N mutations are associated with classical CAH, i.e. SW or SV disease. Prenatal DEX treatment is restricted to families segregating these mutations. Three (P30L, V281L and P453S) of the ten common mutations are associated with NC CAH.

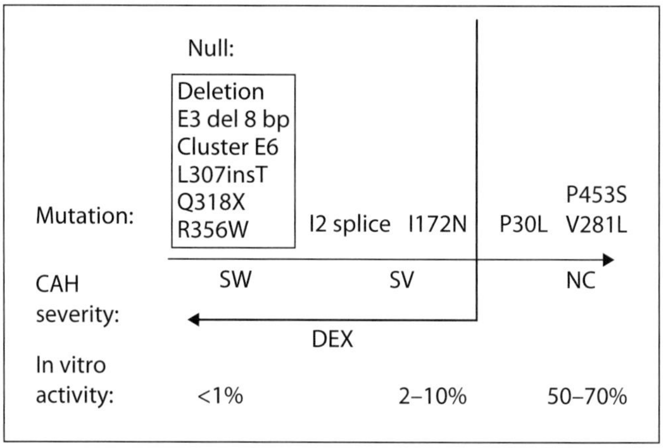

probably reduces the loss of undiagnosed male neonates and leads to earlier diagnosis in both sexes [5–7]. A slightly less severe form of CAH, simple virilizing (SV), is manifested with symptoms similar to those of SW CAH: varying degrees of virilization of external genitalia in female neonates, growth acceleration in childhood, sexual precocity and short stature. However, aldosterone is produced in sufficient amounts to prevent the child from developing salt loss under unstressed conditions. Only a few percent (1–2%) of normal 21OH activity in vitro seem to be enough to prevent SW CAH [8, 9]. The degree of virilization is classified according to the five stages of Prader, where stage 1 is defined as almost normal female external genitalia with minor clitoral enlargement and stage 5 resembles male external genitalia with posterior labial fusion, formation of a urogenital sinus, hypertrophy of the clitoris and a phallic urethra [10]. Individuals with the mildest form of 21OHD, nonclassic (NC) CAH, may escape diagnosis until adulthood when women seek medical care due to hirsutism and fertility problems.

The wide spectrum of symptoms seen in 21OHD is due to differences in enzyme activity, which in turn reflect the underlying mutation(s) in the gene encoding 21-hydroxylase, *CYP21A2,* that maps to the short arm of chromosome 6 (6p21.3) [11]. The locus comprises an active gene, *CYP21A2,* and a nonfunctional pseudogene, *CYP21A1P,* and the great majority of disease-causing mutations in *CYP21A2* alleles are pseudogene-derived sequences transferred to the active gene by microconversions in addition to large gene conversions and gene deletions [12–14]. Most patients with 21OHD (>95%) thus carry one of the ten common pseudogene-derived mutations or a deletion of the entire *CYP21A2* (fig. 2), but an increasing number of novel and rare mutations that are population or family-specific, have been reported (www.imm.ki.se/CYPalleles/cyp21.htm).

There is a good genotype-phenotype correlation in 21OHD. The severity of clinical symptoms is determined by the milder allele carried by the patient. However, there is also a contribution from the other allele, such that a patient with one mildly and one severely affected allele (compound heterozygote) will generally present with a more severe phenotype than one who carries two mild alleles. Recent evidence based on in vitro studies of P450c21, the enzyme that catalyzes adrenal 21-hydroxylation, indicates a good correlation between in vitro enzyme function and disease severity in affected individuals [15–17]. Mutation analysis is particularly useful when the proband is a boy. Families with mutations that give rise to NC disease should never be subjected to prenatal treatment, whereas families carrying the more deleterious mutations may be offered prenatal treatment in a subsequent pregnancy (fig. 2).

Prenatal Diagnosis and Treatment

CYP21A2 genotyping is an invaluable diagnostic tool regardless of whether it is performed by allele-specific PCR of common *CYP21A2* mutations [18] with [19] or without confirmatory linkage analysis or ultimately by direct sequencing of the *CYP21A2* gene. In order to ascertain accurate molecular diagnostics of the fetus using *CYP21A2* genotyping, it is important to analyze both parents as well as the proband ensuring a correct interpretation of the segregation of mutations in this complicated locus. Nowadays, prenatal diagnosis is usually performed on DNA obtained from chorionic villous sampling (CVS) during gestational week 10–12, but obtaining DNA from amniocentesis during gestational weeks 14–16 is of course still an option.

The rational for prenatal treatment of CAH is based on the hypothesis that female in utero virilization can be overcome if adequate fetal adrenal suppression is achieved. The timing of prenatal treatment is also of importance, because the differentiation of the external genital anlage in a male direction, under the influence of testosterone from the fetal testis, is an early embryonic process (gestational week, GW, 6–12). After the early formation of a penis with a male urethra that opens at the glans and fusion of the lateral genital folds to form a scrotum, the external genitalia will continue to grow during the rest of the pregnancy under the influence of androgens. David and Forest [20] described the first successful prenatal treatment of a girl born with 21OHD and thereby defined the outlines of the protocol that is used today for prenatal treatment of CAH (fig. 3). Dexamethasone (DEX) is the drug of choice because it escapes inactivation by placental 11β-hydroxysteroid dehydrogenase and thus suppresses the fetal adrenal cortex more efficiently than hydrocortisone [20]. Suppression of the fetal adrenal cortex will reduce the levels of adrenal androgens and lead to normal female sex differentiation. It has recently been shown that the pituitary-adrenal axis is indeed operating at the early stage when DEX treatment is initiated [21]. Expression of the enzyme 3β-hydroxysteroid dehydrogenase type 2, which is important for the synthesis of cortisol, is detected as early as 8–9 weeks after conception. In addition, at this time point the fetal adrenal is secreting cortisol under

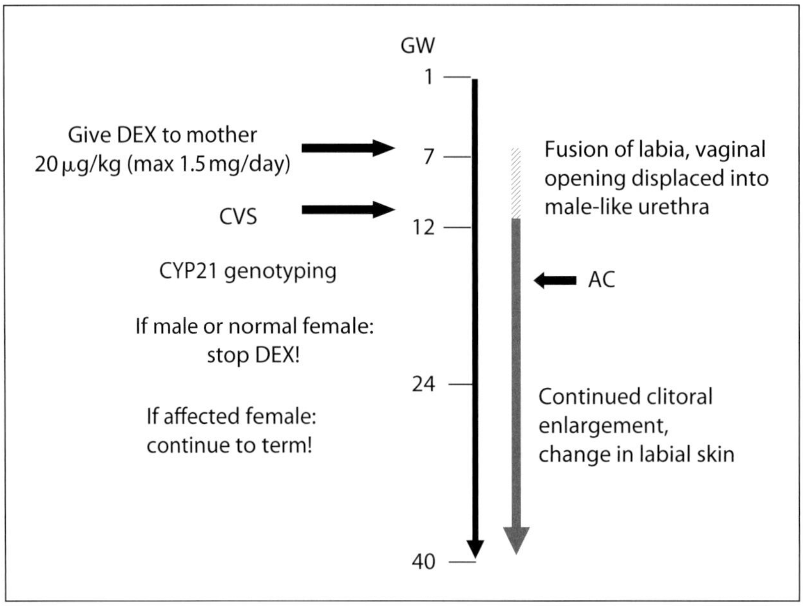

Fig. 3. Protocol for prenatal treatment of CAH. The vertical grey arrow indicates critical periods of androgen action, and shows that the clitoris is sensitive to androgens throughout the entire gestational period. Boys and unaffected girls receive treatment approximately between gestational weeks (GW) 6 and 12–13 if a CVS is performed or between 6 and 16 if amniocentesis (AC) is performed for prenatal diagnostics. CAH-affected girls are treated to term.

the stimulation of ACTH, mediated via the type 2 melanocortin receptor, and the ACTH-secreting corticotrophs are suppressible by DEX [21].

A woman who previously has given birth to a child with severe CAH can be offered treatment with DEX in the 6th–7th gestational week of her pregnancy. The typical dose used is 20 μg/kg body weight/day, divided into three daily doses (max. 1.5 mg/day). A few weeks later, during week 11–12, a CVS is performed to enable genotyping for 21OHD. If the fetus is a male or a healthy female, treatment is stopped; but if it is a female with CAH, treatment is continued until term (fig. 3). In women treated until term, the efficiency of the treatment is monitored with measurements of maternal serum DHEAS (from GW 7) and estriol (from GW 16). Low levels of DHEAS indicate adequate fetal adrenal suppression while a low level of estriol is a marker for maternal adrenal suppression and compliance. Discontinuation of DEX treatment should be tapered, in both short-term and long-term treated women. Due to the recessive mode of inheritance, only 1 out of 8 fetuses will benefit from DEX treatment and 7 out of 8 will be subjected to high doses of GCs unnecessarily during early embryonic life. Today, fetal sex typing (SRY detection) can be performed using PCR amplification of cell-free fetal DNA in maternal blood as early as week 6–9 [22]. Applying this methodology in the genetic diagnostics of CAH would abolish the need for DEX administration and CVS in women carrying male fetuses and only 3 out of 8 fetuses would then be subjected to unnecessary treatment. This method is still only used at a few centers worldwide for diagnostic purposes of 21OHD [23, 24]. Nevertheless, even early sex typing of the fetus still holds the ethical dilemma that a few healthy fetuses will be treated for a short period of time for the sake of benefit of others.

Short-Term Outcome of Prenatal Treatment and Basis for Concerns

The general agreement is that DEX is efficient in reducing or even completely preventing virilization of CAH-affected girls. Treated newborns have consistently been less virilized than their older sisters under the condition that the mother was compliant and that the treatment was initiated in time (mean Prader score 1–2 in treated versus 4 in untreated girls) [25–30]. Most treated girls thus require little or no surgical intervention, which is a great relief for the family and may improve the future quality of life for the child. The effects of excessive androgens on the developing female brain may also be prevented by prenatal treatment. Girls with classic (SW and SV) CAH display more masculine gender-related behavior than healthy girls [31–33], but gender identity does not seem to be affected [34, 35]. Should we then be concerned about plausible side effects of DEX when the positive effects are so obvious? The answer is of course 'yes'. Side effects must be considered in the pregnant woman as well as in the child, as DEX is a very potent synthetic GC that may interfere with the maternal well-being as well as the somatic and psychological development of the growing child. Accumulating evidence tells us that the hormonal and nutritional milieu in utero probably predisposes the child to a range of diseases later in adult life [36, 37]. The fact that 7 out of 8 fetuses receive DEX treatment unnecessarily during the first trimester of pregnancy adds to the concerns. In addition, the panoply of side effects may differ in children treated short-term (GW 7–14) versus long-term (GW 7–40).

Maternal side effects observed during DEX treatment are those normally expected with GC therapy i.e. weight gain, cutaneous striae, edema, sleep disturbances and mood fluctuations. Most of the maternal side effects disappeared after discontinuation of treatment but, in our experience, approximately 50% of the mothers treated during early pregnancy, and all mothers treated until term reported some type of discomfort that could be attributed to DEX [30]. Therefore, it is very important to inform the women in advance about possible side effects in order to achieve better coping and improve compliance. One surprising positive effect was that DEX-treated mothers had significantly less pelvic pain than controls [30]. All studies note that the incidence of hypertension or gestational diabetes does not seem to be increased in DEX-treated women. The incidence of miscarriage is similar to that in the general population or in control groups. While nearly all mothers in the French and New York studies said that they were willing to undergo the treatment again in the event of another pregnancy, about one third of the Swedish mothers reported that they would not [25–30].

Whether prenatal DEX treatment has any adverse effects on the treated fetus has been the subject of intense debate [38, 39]. Therefore, prenatal DEX treatment of children at risk for CAH should only be carried out in centers where proper follow-up of treated cases can be achieved. Excess levels of GC given to the pregnant rat will result in small offspring, but this has not been reported in nonhuman primates and humans, which are considered to be corticosteroid-resistant species [for review, see 40]. Adult corticosteroid-exposed rodent offspring are further characterized by hypertension, hyperinsulinemia, hyperglycemia, hyperactivity of the HPA axis and altered affective behavior, reminiscent of anxiety. On the other hand, early gestation, low-dose DEX treatment of pregnant ewes (20 μg/kg) did not affect blood pressure or renal function in 2 year-old offspring [41].

Most of the prenatally DEX-exposed children at risk of CAH show a normal prenatal and postnatal growth, although several adverse events with failure to thrive have been observed in both short-term- and full-term-treated cases [30]. Long-term studies regarding the metabolic and cardiovascular status are lacking in these children. This should be investigated, as prenatal exposure to GCs programs the fetal HPA axis and may lead to altered susceptibility to metabolic and cardiovascular disease. A single injection of betamethasone given to pregnant sheep during mid-gestation had little effect on HPA function in the offspring at 6 months, but resulted in increased basal and stimulated plasma cortisol levels at 1 year of age [42]. At 2 years of age, offspring of women repeatedly treated with betamethasone during pregnancy had elevated levels of ACTH in response to CRH, but had reduced basal and stimulated cortisol levels [43]. This demonstrates the importance of studying outcome at different points in time during the life course. In the rhesus monkey, prenatal DEX during the last trimester resulted in offspring with elevated levels of basal- and stress-stimulated cortisol at the age of 10 months [44]. In the African vervet monkey, high-dose DEX exposure (120 or 200 μg/kg) from mid-gestation until term resulted in offspring with an exaggerated cortisol response to stress at 12 months of age (prepubertal animals). This reflects an increased drive of the HPA-axis rather than attenuated feedback, as seen in rodents. By the age of 8 months, the animals exhibited hypertension, impaired glucose tolerance and hyperinsulinemia [45]. The observed loss of β-cell mass in 1-year-old animals may lead to overt diabetes in older animals. As in humans, DEX had no effect on birth weight in prenatally treated monkeys, but the length of the femur was reduced demonstrating some fetal growth retardation. However, a DEX dose of 50 μg/kg did not significantly affect any of the outcome measures studied [45]. Little is known about the impact of antenatal GC on human HPA axis function, but a recent report indicates that 30 years after exposure to betamethasone the individuals showed no cardiovascular abnormalities but indication of insulin resistance, particularly in women [46].

Central Nervous System Development and Glucocorticoids

The central nervous system (CNS) is particularly sensitive to a number of teratogenic agents during fetal development. Different aspects of the CNS are affected depending on the developmental phase of the fetus. In early embryonic development, cell proliferation occurs within the neural tube and by week 5 of gestation the basic features of CNS can be identified. After 6 weeks of gestation, the neuroblasts begin to migrate to their permanent locations, where the differentiation begins. Neuronal differentiation includes processes such as development of cell bodies, selective cell death, dendritic and axonal growth, and synaptogenesis [47]. Between the 6th and 12th gestational week many regions of the brain develop, such as the dopaminergic nuclei in the midbrain, the hypothalamic areas and certain areas of the hippocampus, as well as areas of the striatum, amygdala and neocortex [48].

GCs are essential for normal brain development affecting such processes as cell proliferation and neuronal growth and differentiation [49]. Cortisol has a higher affinity for the mineralocorticoid receptor (MR) than for the GC receptor (GR), but synthetic GCs such as DEX bind predominantly to the GR [49]. The MR is expressed primarily in the limbic system, while the GR is present in both subcortical and cortical structures with the highest density found in the hippocampus and the parahippocampal gyrus, the paraventricular nucleus and other hypothalamic nuclei, as well as in the cortex [50]. In primates, the prefrontal cortex is particularly dense in GR [51, 52].

Effects of Excessive Glucocorticoid Levels in Mature, Developing and Fetal Central Nervous System

GC signaling may cause different effects in a mature versus a developing brain, and prenatal exposure to excess GC has other consequences [50]. The effects on adult CNS are often reversible, while effects on a fetus can be organizational, i.e. having an imprinting effect thus 'programming' the behavior and certain physiological functions for long periods of time and even for the entire life span of the individual [53]. It should be noted that behavioral effects including cognition are not necessarily mediated by structural changes in the brain, since GCs are essential for modulation of behavioral adaptation during acute stress, and regulate arousal, alertness, and cognition. Thus, permanent changes in the reactivity of the HPA axis program both cognitive performance and emotional reactions in the affected individual [53, 54].

In adults, the structural effects of excessive endogenous or exogenous GC have consisted in a higher ventricle-brain ratio, enlargement of ventricles and hippocampal athropy, as well as cerebral and cerebellar cortical atrophy [for review, see 55]. A wide range of functional effects have been demonstrated such as negative effects on general intellectual ability, memory, visual and spatial reasoning, concentration, attention and distractibility, as well as impulse inhibition, working memory and other executive functions [55]. Among psychiatric symptoms, excessive GC can increase the risk for depression, anxiety, and agitation [55]. There are few studies on the effects of exogenous administration of GCs on cognitive performance in children, for obvious ethical reasons. Administration of high-dose prednisone to children with chronic asthma caused more severe impairment of verbal memory, as well as more symptoms of anxiety and depression than low doses [56]. Synthetic GCs used in the treatment of CAH were considered to be the most reasonable explanation of the white-matter abnormalities and/or temporal lobe atrophy observed in one third of children [57].

Synthetic GCs are widely used to enhance the maturation of the fetal lung in pregnancies at risk of preterm delivery in order to avoid respiratory distress. A long-term follow-up study of 30-year-old individuals exposed to a single prenatal course of betamethasone showed no effect on neurological, cardiovascular, psychiatric, or cognitive functions [46, 58]. However, more hyperactivity, attention disorders and externalizing problems have been recognized in preschool children who received repeated antenatal betamethasone therapy as compared to children who received a single dose [59]; however, general intellectual ability was not affected [59, 60]. Preterm children with respiratory distress syndrome who received a 1-week course of postnatal DEX therapy were compared to a control group at school age. The DEX group showed significantly poorer motor development and a lower Full-Scale IQ [61]. These differences were not detected in an earlier follow-up of the same cohort at 2 years of age [62], thus illustrating the need for long-term follow-up studies.

The perinatal treatment of preterm children is different from the prenatal treatment of children at risk for CAH, and the effects of GCs may also vary depending on which synthetic GC is chosen [63]. Nonetheless, studies on the treatment of preterm children may provide important information on mechanisms by which synthetic GCs exert their effects. These mechanisms have been studied in more detail in animal models.

Animal Models of Fetal Glucocorticoid Exposure

Experimental data from animals exposed to prenatal corticosteroids have demonstrated adverse effects on somatic development, as well as on cognition and other aspects of behavior. In rats, a range of additional side effects such as low birth weight and hypertension [64], impaired learning and memory functions [65], persistent effects on serotonergic and dopaminergic systems [66] as well as alterations in size and organization of midbrain dopaminergic populations, including a feminization or demasculinization of the three-dimensional cytoarchitecture in males [67]. Reduced exploratory behavior and behavioral inhibition have also been observed in the same species exposed to prenatal DEX treatment either during the entire gestational period or during late gestation [68]. However, the developmental timetable of the fetus differs significantly in rats and humans [48]. Many of the brain areas that develop in rat from mid- to late gestation are formed in humans during the first trimester of pregnancy, such as certain hypothalamic and hippocampal areas [48]. There are also differences in GR distribution in the brain. Primates have more GR in the frontal lobes than the levels described in rodents in which the well-known GC-hippocampus link was originally established [51, 52]. Moreover, rats are considered to be corticosensitive while primates are corticoresistant species [40].

In rhesus monkeys, a single high dose of DEX in late gestation resulted in altered hippocampal architecture. Moreover, multiple injections of DEX over a 24-hour period caused more severe damage on hippocampal neurons than a single injection with the same total dose [69]. In a study of long-term postnatal sequelae, juvenile prenatally DEX-treated rhesus monkeys had significantly higher plasma cortisol at baseline and after stress than controls, as well as a reduction in hippocampal volume [44]. Prenatal administration of DEX to African vervet monkeys was associated with metabolic effects as well as an exaggerated cortisol response to mild stress [45]. As mentioned before, most of the animal models have been designed to imitate the perinatal treatment of premature children or have used high doses of DEX. Little is therefore known about the effects of prenatal DEX treatment as used in CAH.

Psychological Long-Term Follow-Up of Prenatally Treated Children at Risk for Congenital Adrenal Hyperplasia

Comparisons of long-term follow-up studies of prenatal DEX treatment in children at risk for CAH are complicated because different methods of assessment have been used in different age groups (table 1). Two reports focus mainly on behavioral problems and temperament [70; Hirvikoski et al., unpubl. data], while two focus on cognitive and motor development as well as school performance [71, 72].

A pilot study designed as a postal maternal questionnaire survey suggested that prenatally DEX-exposed preschool children showed more shyness, greater emotionality, and less sociability, as well as more internalizing problems [70]. An extended questionnaire study by the same researchers [71] focused on parent-reported cognitive and motor outcome. Data on the children's temperament or behavioral problems were not presented. The other cohort [Hirvikoski et al., unpubl. data], comprised school-age children, but used the same questionnaires when applicable. No between-group differences were observed with the exception of results from the Sociability Scale in Emotionality Activity Shyness Temperament Survey for Children. DEX treated children were rated as slightly more sociable than the controls by their parents. CAH-affected girls and boys (treated with DEX prenatally and hydrocortisone postnatally) were rated similarly to controls.

In the Swedish cohort, increased social anxiety was observed in the CAH-unaffected, short-term-treated group when self-rated, but not in the parental ratings. The modest

Table 1. Results of long-term developmental follow-up studies of prenatal DEX treatment in CAH

Reference	Age	Controls	Neuropsychological tests	Children's self-rating scales	Parental rating scales
Trautman et al. [70]	6 months to 5.5 years (n = 26) Mean = 2.5 ± 1.3	Untreated children at risk for CAH (n = 14)	–	–	Questionnaires measuring development, behavior, as well as temperament No effect on general development (R-DPDQ/ MCDI), or temperament (ITQ/ TTQ/ BSQ/EAS) DEX-treated children had higher scores on Shyness and Emotionality, lower scores on Sociability (EAS), as well as higher Internalizing and Total Problem scores (CBCL)
Meyer-Bahlburg et al. [35, 71]	0–15 months (n = 36)	0–15 months (n = 15)	–	–	Developmental questionnaires: No differences on KIDS No differences on R-PDQ
	15 months to 6 years (n = 89)	15 months to 6 years (n = 126)	–	–	Developmental questionnaires: No differences on CDI No differences on R-PDQ
	6–12 years (n = 44)	6--12 years (n = 162)	–	–	School performance: No differences on CBCL School Scale
	Total n = 174 CAH: 48 (31 girls) Non-CAH: 126	Total n = 313 Untreated children at risk for CAH (195 with CAH)			
Hirvikoski et al. [72]	7–17 years Mean = 10.95 ± 2.33 years (n = 26)	7–17 years Mean = 10.38 ± 2.33 years (n = 35) Age- and sex- matched healthy controls	No differences on IQ (WISC-III), handedness, learning and memory (NEPSY) Short-term treated, CAH-unaffected, children had poorer working memory (Digit Span)	Short-term treated, CAH-unaffected, children rated lower on a questionnaire assessing self-perception of scholastic competence (SPPC), and increased self-rated social anxiety (SASC-R)	School performance: No differences on CBCL School Scale

Table 1. (continued)

Reference	Age	Controls	Neuropsycho-logical tests	Children's self-rating scales	Parental rating scales
Hirvikoski et al. [unpubl.]	7–17 years Mean = 10.95 ± 2.33 years (n = 26)	7–17 years Mean = 10.38 ± 2.33 years (n=35) Age- and sex-matched healthy controls	–	–	Questionnaires measuring behavior and temperament: No differences in behavioral problems (CBCL), inclusive social anxiety (SPAI-C-P) DEX-exposed children were rated as more sociable; no other differences on EAS.

R-DPDQ = The Revised Denver Prescreening Developmental Questionnaire; MCDI = Minnesota Child Development Inventory; CBCL = Child Behavior Checklist; ITQ = Infant Temperament Questionnaire; TTQ = Toddler Temperament Questionnaire; BSQ = Behavioral Style Questionnaire; EAS = EAS Temperament Survey for Children; KIDS = The Kent Infant Development Scale; CDI = The Child Development Inventory; WISC-III = Wechsler Intelligence Scales for Children; NEPSY = Developmental Neuropsychological Assessment; SPPC = The Self-Perception Profile for Children; SASC-R= The Social Anxiety Scale for Children-Revised; SPAI-C-P = Social Phobia and Anxiety Inventory for Children – Parent Report.

parent-child agreement on social anxiety is consistent with previous studies on cross-informant agreement [73]. Given that parents and children focus on different aspects of child psychopathology, multi-source assessment procedures are recommended [73]. The sensitivity of different assessment methods is another important consideration when comparing different studies. Many of the questionnaires are designed to capture clinical problems and may not be sensitive enough to identify subclinical effects. It is important to be aware that subclinical effects though subtle may lead to impairment in everyday life when the child encounters new or challenging tasks.

No differences were observed between the DEX-treated children and controls on the parental ratings of cognitive or motor development in preschool children (table 1) [71]. In addition, there were no differences in parental ratings of school performance in DEX-exposed school-age children (Child Behavior Check List School Scale) as compared to controls [71, 72]. These results were in accord with the normal performance in major cognitive measures such as IQ, learning, and long-term memory observed in the direct neuropsychological assessment of DEX-treated children and controls [72]. However, negative effects on verbal working memory and on the children's perceptions of their scholastic competence were noted [72]. These effects remained significant also when controlling for full-scale IQ and social anxiety. In addition, a nonsignificant statistical trend in a measure of visuospatial working memory was observed.

Working memory is the ability to maintain information for a mental operation during a short period of time [74]. Verbal working memory functions are associated predominantly with effective connectivity in the left prefrontal cortex, whereas visuospatial working memory functions are related to the right prefrontal cortex. Moreover, the catecholaminergic and dopaminergic prefrontal subcortical networks and their connectivity with cerebellar regions are implicated in working memory [75]. Hence, like most cognitive functions and behaviors, working memory emerges from a neural network and not a single brain structure.

Future Perspectives

Long-term follow-up studies of individuals treated with DEX are essential since symptoms may not become obvious until later in life. It should be recognized that prenatal effects on the fetus do not need to affect birth weight in order to have an imprinting effect on metabolic functions or brain development [45]. The risk factors for metabolic and cardiovascular disease such as an aberrant lipid profile, glucose metabolism, kidney function, and blood pressure should be thoroughly investigated. Future studies on individuals treated with DEX should also address issues pertaining to different aspects of the complex regulation and interaction of different parts of the brain. A focus on general intellectual development, learning, and long-term memory is of course essential but other areas of executive function such as response inhibition, working memory, and mental flexibility should also be investigated in greater detail. A direct impact on specific brain structures is not the only factor that may affect cognitive performance, since HPA axis activity is also implicated in the regulation of arousal and cognition. Changes in the neurotransmitter phenotype or white matter may also affect connectivity and hence complex cognitive functions and behaviors emerging from neural networks. Such behavioral patterns as temperament, gender-related behavior, and vulnerability to stress are other important areas for future studies.

Individual differences in vulnerability may complicate the interpretation of data. In animal studies, there have been indications of sex differences with respect to prenatal DEX treatment and behavior [49] as well as HPA function and feedback regulation [76, 77]. Sex differences have not been studied with regard to prenatal treatment in CAH.

Recent studies in primates show that GC programming of the HPA axis does occur [45]. Synthetic GCs given during pregnancy affect the HPA axis at the level of the brain, the pituitary, and the adrenal gland, possibly via modulation of the GR and the MR [54]. The effects may be exerted via facilitated CRF signaling in the amygdala [78]. This may be the link not only between GC exposure and the metabolic syndrome but also between GC exposure and altered stress reactivity. Epigenetics has attracted increasing interest, especially with respect to GC regulation and long-term imprinting effects. Selective methylation and demethylation can permanently alter gene expression

[78]. Permanent changes in transcription factors including methylation changes in the GR promoter can be inherited to the next generation. Effects over generations have been demonstrated in both female and male lines as well as in male offspring, which underscores the transgenerational effect [79]. Whether these effects take place in connection with prenatal DEX treatment in humans remains to be investigated.

References

1 Therrell BL Jr, Berenbaum SA, Manter-Kapanke V, Simmank J, Korman K, Prentice L, Gonzalez J, Gunn S: Results of screening 1.9 million texas newborns for 21-hydroxylase-deficient congenital adrenal hyperplasia. Pediatrics 1998;101:583–590.

2 Auchus RJ: The backdoor pathway to dihydrotestosterone. Trends Endocrinol Metab 2004;15:432–438.

3 Hanley NA, Arlt W: The human fetal adrenal cortex and the window of sexual differentiation. Trends Endocrinol Metab 2006;17:391–397.

4 White PC, Speiser PW: Congenital adrenal hyperplasia due to 21-hydroxylase deficiency. Endocr Rev 2000;21:245–291.

5 Thilen A, Nordenstrom A, Hagenfeldt L, von Dobeln U, Guthenberg C, Larsson A: Benefits of neonatal screening for congenital adrenal hyperplasia (21-hydroxylase deficiency) in sweden. Pediatrics 1998; 101:694.

6 Nordenstrom A, Ahmed S, Jones J, Coleman M, Price DA, Clayton PE, Hall CM: Female preponderance in congenital adrenal hyperplasia due to cyp21 deficiency in england: Implications for neonatal screening. Horm Res 2005;63:22–28.

7 Brosnan PG, Brosnan CA, Kemp SF, Domek DB, Jelley DH, Blackett PR, Riley WJ: Effect of newborn screening for congenital adrenal hyperplasia. Arch Pediatr Adolesc Med 1999;153:1272–1278.

8 Tusie-Luna M-T, Traktman P, White PC: Determination of functional effects of mutations in the ste roid 21-hydroxylase gene (cyp21) using recombinant vaccinia virus. J Biol Chem 1990;265:20916–20922.

9 Lajic S, Robins T, Krone N, Schwarz HP, Wedell A: Cyp21 mutations in simple virilizing congenital adrenal hyperplasia. J Mol Med 2001;79:581–586.

10 Prader A: [genital findings in the female pseudohermaphroditism of the congenital adrenogenital syndrome; morphology, frequency, development and heredity of the different genital forms.]. [german]. Helv Paediatr Acta 1954;9:231–248.

11 White PC, Grossberger D, Onufer BJ, Chaplin DD, New MI, Dupont B, Strominger JL: Two genes encoding steroid 21-hydroxylase are located near the genes encoding the fourth component of complement in man. Proc Natl Acad Sci USA 1985;82: 1089–1093.

12 Morel Y, Murena M, Nicolino M, Carel JC, David M, Forest MG: Correlation between genetic lesions of the cyp21b gene and the clinical forms of congenital adrenal hyperplasia (cah) due to 21-hydroxylase deficiency: Report of a large study of 355 cah chromosomes. Horm Res 1992;37:13.

13 Wedell A: An update on the molecular genetics of congenital adrenal hyperplasia: Diagnostic and therapeutic aspects. J Pediatr Endocrinol Metab 1998; 11:581–589.

14 White PC, Tusie-Luna MT, New MI, Speiser PW: Mutations in steroid 21-hydroxylase (cyp21). Hum Mutat 1994;3:373–378.

15 Lajic S, Clauin S, Robins T, Vexiau P, Blanche H, Bellanne-Chantelot C, Wedell A: Novel mutations in cyp21 detected in individuals with hyperandrogenism. J Clin Endocrinol Metab 2002;87:2824–2829.

16 Robins T, Bellanne-Chantelot C, Barbaro M, Cabrol S, Wedell A, Lajic S: Characterization of novel missense mutations in cyp21 causing congenital adrenal hyperplasia. J Mol Med 2007;85:243–251.

17 Barbaro M, Lajic S, Baldazzi L, Balsamo A, Pirazzoli P, Cicognani A, Wedell A, Cacciari E: Functional analysis of two recurrent amino acid substitutions in the cyp21 gene from italian patients with congenital adrenal hyperplasia. J Clin Endocrinol Metab 2004;89:2402–2407.

18 Wedell A, Thilén A, Ritzén EM, Stengler B, Luthman H: Mutational spectrum of the steroid 21-hydroxylase gene in sweden: Implications for genetic diagnosis and association with disease manifestation. J Clin Endocrinol Metab 1994;78:1145–1152.

19 Mao R, Nelson L, Kates R, Miller CE, Donaldson DL, Tang W, Ward K: Prenatal diagnosis of 21-hydroxylase deficiency caused by gene conversion and rearrangements: Pitfalls and molecular diagnostic solutions. Prenat Diagn 2002;22:1171–1176.

20 David M, Forest MG: Prenatal treatment of congenital adrenal hyperplasia resulting from 21-hydroxylase deficiency. J Pediatr 1984;105:799–803.

21 Goto M, Piper Hanley K, Marcos J, Wood PJ, Wright S, Postle AD, Cameron IT, Mason JI, Wilson DI, Hanley NA: In humans, early cortisol biosynthesis provides a mechanism to safeguard female sexual development. J Clin Invest 2006;116:953–960.

22 Bischoff FZ, Lewis DE, Simpson JL: Cell-free fetal DNA in maternal blood: Kinetics, source and structure. Hum Reprod Update 2005;11:59–67.

23 Rijnders RJ, van der Schoot CE, Bossers B, de Vroede MA, Christiaens GC: Fetal sex determination from maternal plasma in pregnancies at risk for congenital adrenal hyperplasia. Obstet Gynecol 2001;98:374–378.

24 Morel Y, Tardy V, Costa JM, Forest MG, David M: [21 hydroxylase deficiency: New strategies emerging from molecular studies]. Ann Endocrinol (Paris) 2003;64:456–470.

25 Forest MG, David M, Morel Y: Prenatal diagnosis and treatment of 21-hydroxylase deficiency. J Steroid Biochem Mol Biol 1993;45:75–82.

26 Forest MG, Dörr HG: Prenatal therapy in congenital adrenal hyperplasia due to 21-hydroxylase deficiency: Retrospective follow-up study of 253 treated pregnancies in 215 families. Endocrinologist 2003;13:252–259.

27 Mercado AB, Wilson RC, Cheng KC, Wei J-Q, New MI: Prenatal treatment and diagnosis of congenital adrenal hyperplasia owing to steroid 21-hydroxylase deficiency. J Clin Endocrinol Metab 1995;80:2014–2020.

28 Carlson AD, Obeid JS, Kannelopoulou N, Wilson RC, New MI: Congenital adrenal hyperplasia: Update on prenatal diagnosis and treatment. J Steroid Biochem Mol Biol 1999;69:19–29.

29 New MI, Carlson A, Obeid J, Marshall I, Cabrera MS, Goseco A, Lin-Su K, Putnam AS, Wei JQ, Wilson RC: Prenatal diagnosis for congenital adrenal hyperplasia in 532 pregnancies. J Clin Endocrinol Metab 2001;86:5651–5657.

30 Lajic S, Wedell A, Bui TH, Ritzen EM, Holst M: Long-term somatic follow-up of prenatally treated children with congenital adrenal hyperplasia. J Clin Endocrinol Metab 1998;83:3872–3880.

31 Berenbaum SA, Duck SC, Bryk K: Behavioral effects of prenatal versus postnatal androgen excess in children with 21-hydroxylase-deficient congenital adrenal hyperplasia. J Clin Endocrinol Metab 2000;85:727–733.

32 Servin A, Nordenstrom A, Larsson A, Bohlin G: Prenatal androgens and gender-typed behavior: A study of girls with mild and severe forms of congenital adrenal hyperplasia. Dev Psychol 2003;39:440–450.

33 Hall CM, Jones JA, Meyer-Bahlburg HF, Dolezal C, Coleman M, Foster P, Price DA, Clayton PE: Behavioral and physical masculinization are related to genotype in girls with congenital adrenal hyperplasia. J Clin Endocrinol Metab 2004;89:419–424.

34 Berenbaum SA, Bailey JM: Effects on gender identity of prenatal androgens and genital appearance: Evidence from girls with congenital adrenal hyperplasia. J Clin Endocrinol Metab 2003;88:1102–1106.

35 Meyer-Bahlburg HF, Dolezal C, Baker SW, Carlson AD, Obeid JS, New MI: Prenatal androgenization affects gender-related behavior but not gender identity in 5–12-year-old girls with congenital adrenal hyperplasia. Arch Sex Behav 2004;33:97–104.

36 Barker DJ, Eriksson JG, Forsen T, Osmond C: Fetal origins of adult disease: Strength of effects and biological basis. Int J Epidemiol 2002;31:1235–1239.

37 Seckl JR, Holmes MC: Mechanisms of disease: Glucocorticoids, their placental metabolism and fetal 'programming' of adult pathophysiology. Nat Clin Pract Endocrinol Metab 2007;3:479–488.

38 Seckl JR, Miller WL: How safe is long-term prenatal glucocorticoid treatment? JAMA 1997;277:1077–1079.

39 Miller WL: Dexamethasone treatment of congenital adrenal hyperplasia in utero: An experimental therapy of unproven safety. J Urol 1999;162:537–540.

40 Seckl JR: Prenatal glucocorticoids and long-term programming. Eur J Endocrinol 2004;151(suppl 3):U49–U62.

41 Dodic M, Tersteeg M, Jefferies A, Wintour EM, Moritz K: Prolonged low-dose dexamethasone treatment, in early gestation, does not alter blood pressure or renal function in adult sheep. J Endocrinol 2003;179:275–280.

42 Sloboda DM, Moss TJ, Gurrin LC, Newnham JP, Challis JR: The effect of prenatal betamethasone administration on postnatal ovine hypothalamic-pituitary-adrenal function. J Endocrinol 2002;172:71–81.

43 Sloboda DM, Moss TJ, Li S, Doherty D, Nitsos I, Challis JR, Newnham JP: Prenatal betamethasone exposure results in pituitary-adrenal hyporesponsiveness in adult sheep. Am J Physiol Endocrinol Metab 2007;292:E61–E70.

44 Uno H, Eisele S, Sakai A, Shelton S, Baker E, DeJesus O, Holden J: Neurotoxicity of glucocorticoids in the primate brain. Horm Behav 1994;28:336–348.

45 de Vries A, Holmes MC, Heijnis A, Seier JV, Heerden J, Louw J, Wolfe-Coote S, Meaney MJ, Levitt NS, Seckl JR: Prenatal dexamethasone exposure induces changes in nonhuman primate offspring cardiometabolic and hypothalamic-pituitary-adrenal axis function. J Clin Invest 2007;117:1058–1067.

46 Dalziel SR, Walker NK, Parag V, Mantell C, Rea HH, Rodgers A, Harding JE: Cardiovascular risk factors after antenatal exposure to betamethasone: 30-year follow-up of a randomised controlled trial. Lancet 2005;365:1856–1862.
47 Anderson V, Northam E, Hendy J, Wrennall J: Cerebral development. Developmental neuropsychology: A clinical approach. Hove, Psychology Press, 2001, pp 39–68.
48 Bayer SA, Altman J, Russo RJ, Zhang X: Timetables of neurogenesis in the human brain based on experimentally determined patterns in the rat. Neurotoxicology 1993;14:83–144.
49 Matthews SG: Antenatal glucocorticoids and the developing brain: Mechanisms of action. Semin Neonatol 2001;6:309–317.
50 McEwen B, Chao H, Spencer R, Brinton R, Macisaac L, Harrelson A: Corticosteroid receptors in brain: Relationship of receptors to effects in stress and aging. Ann NY Acad Sci 1987;512:394–401.
51 Patel PD, Lopez JF, Lyons DM, Burke S, Wallace M, Schatzberg AF: Glucocorticoid and mineralocorticoid receptor mrna expression in squirrel monkey brain. J Psychiatr Res 2000;34:383–392.
52 Sanchez MM, Young LJ, Plotsky PM, Insel TR: Distribution of corticosteroid receptors in the rhesus brain: Relative absence of glucocorticoid receptors in the hippocampal formation. J Neurosci 2000; 20:4657–4668.
53 Charmandari E, Kino T, Souvatzoglou E, Chrousos GP: Pediatric stress: Hormonal mediators and human development. Horm Res 2003;59:161–179.
54 Kapoor A, Petropoulos S, Matthews SG: Fetal programming of hypothalamic-pituitary-adrenal (hpa) axis function and behavior by synthetic glucocorticoids. Brain Res Rev 2008;57:586–595.
55 Belanoff JK, Gross K, Yager A, Schatzberg AF: Corticosteroids and cognition. J Psychiatr Res 2001;35: 127–145.
56 Bender BG, Lerner JA, Poland JE: Association between corticosteroids and psychologic change in hospitalized asthmatic children. Ann Allergy 1991; 66:414–419.
57 Nass R, Heier L, Moshang T, Oberfield S, George A, New MI, Speiser PW: Magnetic resonance imaging in the congenital adrenal hyperplasia population: Increased frequency of white-matter abnormalities and temporal lobe atrophy. J Child Neurol 1997;12: 181–186.
58 Dalziel SR, Lim VK, Lambert A, McCarthy D, Parag V, Rodgers A, Harding JE: Antenatal exposure to betamethasone: Psychological functioning and health related quality of life 31 years after inclusion in randomised controlled trial. BMJ 2005;331:665.
59 French NP, Hagan R, Evans SF, Mullan A, Newnham JP: Repeated antenatal corticosteroids: Effects on cerebral palsy and childhood behavior. Am J Obstet Gynecol 2004;190:588–595.
60 Wapner RJ, Sorokin Y, Mele L, Johnson F, Dudley DJ, Spong CY, Peaceman AM, Leveno KJ, Malone F, Caritis SN, Mercer B, Harper M, Rouse DJ, Thorp JM, Ramin S, Carpenter MW, Gabbe SG: Long-term outcomes after repeat doses of antenatal corticosteroids. N Engl J Med 2007;357:1190–1198.
61 Yeh TF, Lin YJ, Lin HC, Huang CC, Hsieh WS, Lin CH, Tsai CH: Outcomes at school age after postnatal dexamethasone therapy for lung disease of prematurity. N Engl J Med 2004;350:1304–1313.
62 Yeh TF, Lin YJ, Huang CC, Chen YJ, Lin CH, Lin HC, Hsieh WS, Lien YJ: Early dexamethasone therapy in preterm infants: A follow-up study. Pediatrics 1998;101:E7.
63 Baud O, Sola A: Corticosteroids in perinatal medicine: How to improve outcomes without affecting the developing brain? Semin Fetal Neonatal Med 2007;12:273–279.
64 Celsi G, Kistner A, Aizman R, Eklof AC, Ceccatelli S, de Santiago A, Jacobson SH: Prenatal dexamethasone causes oligonephronia, sodium retention, and higher blood pressure in the offspring. Pediatr Res 1998;44:317–322.
65 Emgard M, Paradisi M, Pirondi S, Fernandez M, Giardino L, Calza L: Prenatal glucocorticoid exposure affects learning and vulnerability of cholinergic neurons. Neurobiol Aging 2007;28:112–121.
66 Slotkin TA, Kreider ML, Tate CA, Seidler FJ: Critical prenatal and postnatal periods for persistent effects of dexamethasone on serotonergic and dopaminergic systems. Neuropsychopharmacology 2006; 31:904–911.
67 McArthur S, McHale E, Gillies GE: The size and distribution of midbrain dopaminergic populations are permanently altered by perinatal glucocorticoid exposure in a sex- region- and time-specific manner. Neuropsychopharmacology 2007;32:1462–1476.
68 Welberg LA, Seckl JR, Holmes MC: Prenatal glucocorticoid programming of brain corticosteroid receptors and corticotrophin-releasing hormone: Possible implications for behaviour. Neuroscience 2001;104: 71–79.
69 Uno H, Lohmiller L, Thieme C, Kemnitz JW, Engle MJ, Roecker EB, Farrell PM: Brain damage induced by prenatal exposure to dexamethasone in fetal rhesus macaques. I. Hippocampus. Brain Res Dev Brain Res 1990;53:157–167.
70 Trautman PD, Meyer-Bahlburg HF, Postelnek J, New MI: Effects of early prenatal dexamethasone on the cognitive and behavioral development of young children: Results of a pilot study. Psychoneuroendocrinology 1995;20:439–449.

71 Meyer-Bahlburg HF, Dolezal C, Baker SW, Carlson AD, Obeid JS, New MI: Cognitive and motor development of children with and without congenital adrenal hyperplasia after early-prenatal dexamethasone. J Clin Endocrinol Metab 2004;89:610–614.
72 Hirvikoski T, Nordenstrom A, Lindholm T, Lindblad F, Ritzen EM, Wedell A, Lajic S: Cognitive functions in children at risk for congenital adrenal hyperplasia treated prenatally with dexamethasone. J Clin Endocrinol Metab 2007;92:542–548.
73 Achenbach TM, McConaughy SH, Howell CT: Child/adolescent behavioral and emotional problems: Implications of cross-informant correlations for situational specificity. Psychol Bull 1987;101:213–232.
74 Baddeley A: Working memory. Oxford, Clarendon Press, 1986.
75 Arnsten AF: Fundamentals of attention-deficit/hyperactivity disorder: Circuits and pathways. J Clin Psychiatry 2006;67(suppl 8):7–12.
76 Liu L, Li A, Matthews SG: Maternal glucocorticoid treatment programs hpa regulation in adult offspring: Sex-specific effects. Am J Physiol Endocrinol Metab 2001;280:E729–E739.
77 Banjanin S, Kapoor A, Matthews SG: Prenatal glucocorticoid exposure alters hypothalamic-pituitary-adrenal function and blood pressure in mature male guinea pigs. J Physiol 2004;558:305–318.
78 Meaney MJ, Szyf M, Seckl JR: Epigenetic mechanisms of perinatal programming of hypothalamic-pituitary-adrenal function and health. Trends Mol Med 2007;13:269–277.
79 Drake AJ, Walker BR, Seckl JR: Intergenerational consequences of fetal programming by in utero exposure to glucocorticoids in rats. Am J Physiol Regul Integr Comp Physiol 2005;288:R34–R38.

Svetlana Lajic, MD, PhD
Department of Molecular Medicine and Surgery
Centre for Molecular Medicine L8:02, Karolinska University Hospital
SE–171 76 Stockholm (Sweden)
Tel. +46 8 5177 3922, Fax +46 8 5177 3620, E-Mail Svetlana.Lajic@ki.se

Flück CE, Miller WL (eds): Disorders of the Human Adrenal Cortex.
Endocr Dev. Basel, Karger, 2008, vol 13, pp 99–116

Adrenocorticotropin Resistance Syndromes

Sadani N. Cooray · Li Chan · Lou Metherell · Helen Storr · Adrian J.L. Clark

Centre for Endocrinology, William Harvey Research Institute, Barts and the London, London, UK

Abstract

Familial glucocorticoid deficiency (FGD) and triple A syndrome belong to a rare group of autosomal recessive disorders characterized by adrenocorticotropin (ACTH) insensitivity. Unlike triple A syndrome which presents a range of clinical features, FGD is solely characterized by glucocorticoid deficiency. ACTH regulates steroid biosynthesis in the adrenal cortex by exerting its effects via the ACTH receptor (melanocortin-2 receptor, MC2R). In FGD, mutations in the MC2R account for only approximately 25% of cases (FGD type 1). The inability to express a functional MC2R in non-adrenal cell lines had implied the presence of an adrenal specific accessory factor(s), essential for MC2R expression. More recently, this factor was identified as melanocortin receptor accessory protein (MRAP). Mutations in MRAP account for ~20% of cases (FGD type 2). Like the receptor activity-modifying proteins (RAMPs) and receptor transporter proteins (RTPs), which are well-characterized accessory proteins for G-protein-coupled receptors (GPCRs), MRAP is a small single transmembrane domain protein. MRAP is essential for the functional expression of the MC2R. About 55% of FGD cases have no identifiable gene defect, implying the involvement of additional genes. This chapter briefly describes the clinical and biochemical features of ACTH resistance syndromes. However, we will focus on the recent progress made towards understanding the molecular defect underlying these conditions, in particular the interaction of MC2R and MRAP.

Melanocortin Peptides

The melanocortins are a group of pituitary peptide hormones that include adrenocorticotropin (ACTH) and melanocyte-stimulating hormones (α-MSH, β-MSH, γ-MSH and δ-MSH). The term melanocortin is assigned to these peptides because of their ability to stimulate eumelanogenesis in the melanocyte and/or to stimulate steroid production in the adrenocortical cell.

In the 1950s, ACTH and α-MSH were purified from anterior pituitary extracts and their protein sequences determined [1, 2]. α-MSH was found to be part of the ACTH sequence. The sequence of β-MSH was different, but shared a common heptapeptide

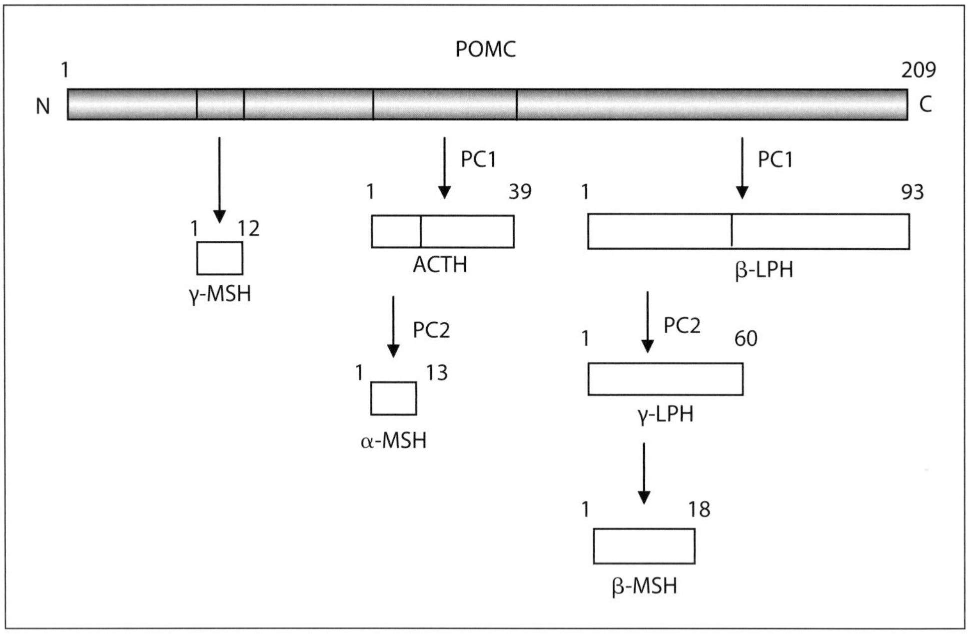

Fig. 1. Processing of POMC in the pituitary to smaller peptides. The numbers show the length of the amino acid sequence of the peptides. The enzymatic cleavage by the prohormone convertase PC1 and PC2 is shown.

core with ACTH and α-MSH and was contained within the sequence of the β/γ-lipotropin. These findings led to the hypothesis that the shorter peptide forms are derived from a longer precursor. In 1979, Nakanishi et al. [3] demonstrated that all these peptides were derived from the precursor protein proopiomelanocortin (POMC). Furthermore, they showed that POMC was processed into another MSH-like peptide termed γ-MSH. Figure 1 shows the proteolytic cleavage of POMC by prohormone convertase enzymes PC1 and PC2 into the various melanocortin peptides. The structural feature that characterizes all MSH sequences and ACTH is the core tetrapeptide His-Phe-Arg-Trp (HFRW) that is crucial for these peptides to interact with their receptors (melanocortin receptors, MCRs) (table 1).

The melanocortin peptides derived from POMC differ depending on where they are produced. POMC is primarily synthesized by corticotroph cells of the anterior pituitary gland that secrete ACTH, which in turn regulates the secretion of glucocorticoids. Most mammals, including the human fetus, also produce POMC by melanotroph cells of the intermediate lobe of the pituitary gland that secrete α-MSH to regulate the production of melanin. However, due to its minimal development in human adults, the intermediate lobe of the pituitary gland is believed to be vestigial. POMC is also expressed by neurons in the arcuate nucleus of the hypothalamus and

Table 1. Alignment of the melanocortin peptide sequences: HFRW (His-Phe-Arg-Trp) depicts the tetrapeptide sequence that is essential for receptor binding

ACTH	H2N-SYSMEHFRWGKPVGKKRRVKVYPNGAEDESEAAFPLEF-OH
α-MSH	N-Acetyl-SYSMEHFRWGKPV-NH2
β-MSH	H2N-AEKKDEGRYRMEHFRWGSPPKD-OH
γ-MSH	H2N-YVMGHFRWDRF-OH

in a range of peripheral tissues. Apart from the classical functions of melanogenesis and steroidogenesis, they exert a range of diverse effects including weight and energy homeostasis [4].

MCR Family

The melanocortin peptides exert their numerous biological effects by activating receptors (MCRs) that belong to the rhodopsin/β_2-adrenergic-like family A of the G-protein-coupled receptor (GPCR) superfamily. The MSH receptor was the first to be cloned in 1992 by two independent groups [5, 6]. To date five MCRs have been identified, termed MC1R–MC5R according to the order in which they were cloned. MCRs exhibit the common structural characteristics of the GPCR family, consisting of seven α-helical transmembrane domains that are interconnected by six alternating extra- and intracellular peptide loops. The MCRs are among the smallest GPCRs known, with short N- and C-terminal ends. They are all positively coupled to adenylyl cyclase and mediate their effects primarily by activating the cAMP-dependent signalling pathway [7, 8]. MCRs share many features common to other GPCRs, including potential N-glycosylation sites in their N-terminal domains, recognition sites for protein kinase A and protein kinase C and conserved cysteines in their C-terminal ends [9–11]. Each of the five MCRs has distinct ligand specificity, tissue distribution and biological significance (summarized in table 2).

Melanocortin-2 Receptor/ACTH Receptor

The melanocortin-2 receptor (MC2R) is unique among the MCRs because it is only activated by ACTH. Activation of the MC2R results in an increase in cAMP and PKA activity, which is essential for promoting the expression of steroidogenic enzymes. Haynes [12] in 1958 established for the first time the ability of ACTH to stimulate the production of cAMP in the adrenal cortex. Subsequently, ligand binding studies carried out by Lefkowitz et al. [13] confirmed the idea that cAMP production by ACTH

Table 2. Biological distribution and functions of the MCR family

Receptor	Ligand preference	Site of expression	Main function
MC1R	α-MSH > ACTH > β-MSH > γ-MSH	melanocytes, melanoma, macrophage, brain	pigmentation, anti-inflammatory, antipyretic
MC2R	ACTH only	adrenal cortex	steroidogenesis
MC3R	γ-MSH > α-MSH/β-MSH/ACTH	brain, placenta, pancreas, stomach	energy homeostasis
MC4R	α-MSH > β-MSH/ACTH > γ-MSH	brain, spinal cord	appetite regulation, energy homeostasis, erectile dysfunction
MC5R	α-MSH > β-MSH/ACTH > γ-MSH	widely expressed in tissues including adrenal gland, skin, muscle, kidney, lung, lymphocytes and exocrine glands	sebaceous secretion control, immunoregulatory functions

was mediated through a specific cell surface receptor now referred to as MC2R or ACTH receptor.

The human MC2R gene was cloned in 1992 by Mountjoy et al. [6]. Its identity was based on homology with the MC1R and its specific expression in the adrenal cortex. The mouse MC2R was cloned in 1995 by Cammas et al. [14] using a strategy of low stringency PCR and mouse genomic DNA library screening. The mouse MC2R gene was found to be expressed in both mouse adrenals and Y1 mouse adrenocortical cells. In situ hybridization studies using mouse adrenals demonstrated that MC2R is found predominantly in the zona glomerulosa and fasciculata with very low expression in the zona reticularis and adrenal medulla [15].

The human MC2R gene encodes a protein of 297 amino acids while the mouse MC2R encodes a protein of 296 amino acids, making MC2R the smallest among all known GPCRs. The human and mouse MC2R proteins show 84% amino acid identity. Human MC2R has a predicted molecular weight of 33 kDa in its unmodified form and has two N-linked glycosylation sites at the amino terminus. Prior to the cloning of MC2R, several groups including Hofmann et al. [16] attempted to purify MC2R using conventional biochemical methods. They demonstrated that MC2R was expressed as a 43-kDa protein at the cell surface, which is compatible with the molecular weight of the glycosylated form of the receptor.

Expression of MC2R in Heterologous Cells

The functional characterization of the human MC2R in heterologous cells has been extremely challenging due to lack of expression of the transfected receptor. For this reason there was initially some doubt as to whether the MC2R was indeed the receptor for ACTH. Several groups attempted to express and functionally characterize the MC2R in a number of cell lines including HEK 293 and CHO cells, and failed. Weber et al. [17] used a highly optimized protocol to transfect COS-7 cells with the MC2R and only managed to detect faint expression of the receptor. This result was further complicated by the generation of cAMP by an endogenous MCRs responding to ACTH. Similarly characterization of the MC2R transfected into the Cloudman M3 melanoma cell line was confounded due to the presence of MC1R that is endogenously expressed in these cells [18].

Subsequently the Y1 mouse adrenocortical cell line that expresses MC2R alone was used in order to investigate the binding of ACTH and MSH peptides using radiolabelled ACTH. The cells showed high affinity binding to ^{125}I ACTH but could not bind any of the other melanocortin peptides. However, when the other four MCRs (MC1R, MC3R, MC4R and MC5R) were transfected into Y1 cells binding was identified with ^{125}I NDP-MSH (Nle4, D-Phe7, a potent and enzymatically stable analogue of α-MSH) for each of the receptors with no specific binding in untransfected cells [19], demonstrating that MC2R is the ACTH receptor and only binds ACTH.

The use of cell lines Y6 and OS3 (sister cell lines of Y1 cells), which fail to express endogenous MC2R, allowed the functional characterization of the MC2R using radioligand binding and cAMP assays, since there was satisfactory expression of a transfected MC2R with no other interfering MCRs [20, 21].

MC2R Signalling and Adrenal Steroidogenesis

Steroid biosynthesis in the adrenal cortex is under the control of ACTH acting via the MC2R. The importance of ACTH in adrenal steroidogenesis was made evident by experiments carried out as early as 1969 when it was shown that the hypophysectomy of rats led to a substantial decline in the activities of steroid hydroxylases and that treatment of hypophysectomized animals with pharmacological doses of ACTH restored these enzymatic reactions [22]. ACTH regulates steroidogenesis by two pathways [23–25]. First, an acute response involves the mobilization of cholesterol from intracellular stores into the mitochondrion by the steroidogenic acute regulatory protein (StAR), where the conversion of cholesterol to pregnenolone is catalyzed by P450 side chain cleavage enzyme (CYP11A1). This is the first step in all steroidogenic pathways (fig. 2).

Second, by a slower response, ACTH regulates the transcription of genes that encode enzymes of the steroidogenic pathway. These include four steroid hydroxylases

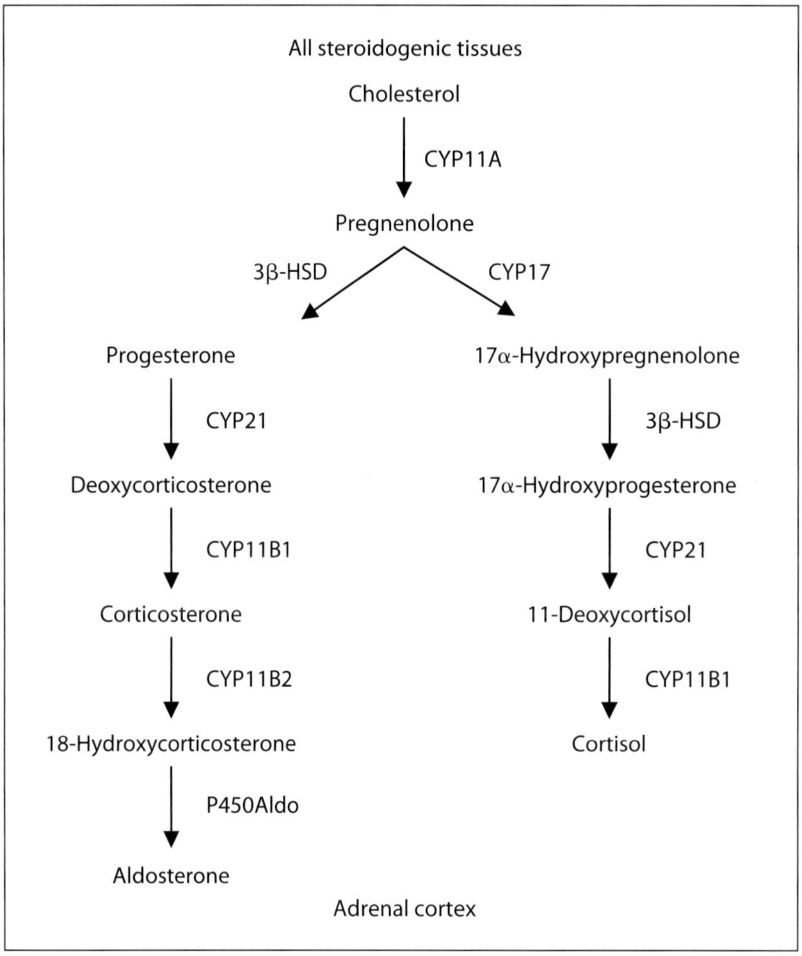

Fig. 2. Schematic representation of the steroidogenic pathways: ACTH regulates the expression of the adrenal steroidogenic enzyme genes, which catalyze reactions leading to glucocorticoid and mineralocorticoid synthesis.

of the cytochrome P450 (CYP) superfamily, namely, 17α-hydroxylase (CYP17), 21-hydroxylase (CYP21), 11β-hydroxylase (CYP11 B1) and aldosterone synthase (CYP11 B2). Both rapid and slow responses are mediated by a cAMP-dependent pathway subsequent to the activation of the MC2R. Activation of protein kinase A (PKA) results in both acute and chronic effects. During the acute response, PKA phosphorylates hormone-sensitive lipase which then catalyzes the conversion of cholesterol esters to free cholesterol [26], whilst in the chronic response PKA directs transcription of steroidogenic genes [23, 25].

ACTH Insensitivity Syndromes

ACTH insensitivity syndromes are a group of rare hereditary disorders resulting in adrenal unresponsiveness to ACTH. If untreated, patients with these disorders face life-threatening illnesses that can ultimately result in death. Triple A syndrome and familial glucocorticoid deficiency (FGD) are both inherited forms of ACTH insensitivity that result in glucocorticoid deficiency due to the inability of ACTH to stimulate steroidogenesis in the adrenal cortex.

Triple A Syndrome

Triple A syndrome, first described in 1978 by Allgrove et al. [27], consists of a triad of alacrima (deficient tear production), achalasia (swallowing problems) and adrenal failure. Isolated glucocorticoid deficiency is seen in approximately 80% of patients with additional mineralocorticoid deficiency noted in 15%. Not all patients have the main features characterizing this syndrome; for example, alacrima is present in 90% of cases, whilst achalasia is present in ~85% of patients. Even when present, the symptoms and signs may be subtle and often require further investigation, such as a Schirmer test for tear production or a barium swallow for a diagnosis of achalasia. Since the first description, a wide range of associated progressive neurological defects have also been reported. These include motor, sensory and autonomic neuropathies, mental retardation and dementia, optic atrophy and sensorineural deafness [28, 29]. Histology of the adrenal glands taken from patients who had succumbed to this disease showed a preserved zona glomerulosa with an atrophic zona fasciculata and zona reticularis.

It was previously proposed that defects in the MC2R might be associated with triple A, however, several groups failed to identify any mutations in the receptor [30, 31]. Since then the syndrome has been mapped to chromosome 12q13 [32] and defects in the AAAS gene have been identified in patients with triple A [33–35]. These mutations are located throughout the gene but only account for about 90% of triple A, making it a genetically heterogeneous disorder. The AAAS gene encodes a protein named ALADIN, alacrima-achalasia-adrenal insufficiency neurological disorder, which is a WD repeat protein [34].

The presence of WD repeats (a β-propeller structure) suggests that ALADIN is likely to be involved in protein-protein interactions and supports the formation of multimolecular complexes [36–38]. ALADIN was recently identified as one of the proteins comprising the nuclear pore complex (NPC) [39, 40] and was localized to the cytoplasmic face of the NPC [41]. ALADIN mutations result in a functional rather than a structural NPC defect and triple A syndrome may result from a tissue-specific failure of nuclear pore function [42]. Many of the naturally occurring missense, nonsense and frameshift *AAAS* mutations result in a failure of ALADIN to localize to the NPC [42–44].

The *AAAS* knockout mouse (Aaas−/−) does not provide a good disease model [45]. Although Aaas−/− mice weighed less than the wild-type (WT) mice and the Aaas−/− females were unfertile, they were externally indistinguishable from their WT littermates and histological analysis of the adrenal gland, oesophagus, ovary, pituitary gland, testes, and peripheral nerves was normal [45]. The pathogenesis of triple A syndrome may involve oxidative stress [46]. Hirano et al. [46] demonstrated that nuclear import was impaired in affected patients' cells and, as a consequence, there was increased cellular susceptibility to oxidative stress, i.e. DNA damage and cell death. Nuclear import in general was impaired and the accumulation of aprataxin (APTX), a DNA single-strand break (SSBs) repair protein, and DNA ligase 1 was found to be reduced in cultured patient fibroblasts.

Triple A syndrome is an extremely rare genetic disease. The combination of clinical manifestations is unique and includes a range of phenotypic abnormalities that can be highly heterogeneous even within affected families. The current treatments available for the features of the disorder are aimed at symptom relief rather than prevention or cure [27, 28, 47]. The eventual determination of the exact functional role(s) of the ALADIN protein will provide a fascinating insight into the pathogenesis of the triple A syndrome and may suggest future therapeutic interventions.

Familial Glucocorticoid Deficiency

FGD, also known as hereditary unresponsiveness to ACTH or isolated glucocorticoid deficiency, is a rare autosomal recessive disorder that is characterized by severe cortisol deficiency, high plasma ACTH levels and typically, a well-preserved renin-angiotensin-aldosterone axis and hence normal mineralocorticoid levels [28].

This syndrome was initially described by Shepard et al. [48] in 1959 when they reported the disease in two sisters. Further cases demonstrated the familial nature of the disease and the onset of symptoms in the neonatal period [49–51]. The observation that the parents of these patients had no adrenal irregularities combined with parental consanguinity in some cases led to the suggestion that FGD was an autosomal recessive disorder that was most prevalent in consanguineous families [52].

Clinical and Histological Features

Patients with FGD usually present with symptoms of glucocorticoid deficiency during the neonatal period or by early childhood. Symptoms include hypoglycaemic seizures, recurrent infections, failure to thrive, collapse and coma. Neonatal hepatitis, a known complication of hypocortisolaemia, has also been described in FGD [53]. Biochemical investigations typically reveal extremely high plasma ACTH levels paired with low or undetectable cortisol levels. Plasma ACTH concentrations are often in excess of

1,000 pg/ml (normal range <80 pg/ml by RIA; <50 pg/ml by IRMA) and typically remain elevated despite treatment with fairly high doses of hydrocortisone. High ACTH levels acting on MC1R in cutaneous melanocytes lead to excessive skin pigmentation, a feature usually seen by 6 weeks of life in patients with FGD.

FGD is characterized by the absence of mineralocorticoid deficiency, although mild derangements in the renin-angiotensin-aldosterone axis at the time of presentation are well established. Such transient changes occur at times of acute illness and can partly explain how patients may be misdiagnosed [54]. As the production of aldosterone from the zona glomerulosa is principally controlled by angiotensin II, changes in mineralocorticoid secretion would not be expected in FGD. However, evidence that ACTH may play a direct role in aldosterone regulation comes from several observations. First, MC2R mRNA is present in the zona glomerulosa [55]. Second, administration of synthetic ACTH to normal human subjects results in a rise in plasma aldosterone levels [56] and finally, disorganization of the zona glomerulosa is seen on adrenal histology of deceased FGD patients [28]. Furthermore, the recently reported phenotype of the MC2R knockout mouse supports the notion that ACTH/MC2R signalling is important in the regulation of aldosterone production. Knockout mice had significantly lower serum aldosterone levels compared to WT mice, as well as increased expression of the angiotensin receptor 1b suggesting the need for compensation in the renin-angiotensin-aldosterone system in the complete absence of ACTH signalling [57]. In humans the physiological significance remains undetermined.

MC2R Defects and FGD

Cloning of the human MC2R in 1992 [6] provided an opportunity to investigate the hypothesis that defects in the MC2R could cause FGD. In 1993, Clark et al. [58] reported a homozygous missense point mutation converting Ser74 to Ile (S74I) in two affected siblings. Subsequently, a number of different missense and nonsense mutations were reported [32, 59–63]. Since then more than 35 mutations have been identified in the MC2R that are scattered throughout the receptor (fig. 3). The S74I mutation is the most prevalent mutation described to date.

The functional consequences of these mutations include loss of ligand binding, truncation of the receptor, disruptions of the transmembrane domain and loss of signal transduction [20, 31, 61].

Genetic Heterogeneity of FGD

DNA sequence analysis of all patients with FGD has demonstrated that only about 25% of FGD is caused by mutations in the MC2R and, in many cases, there was no

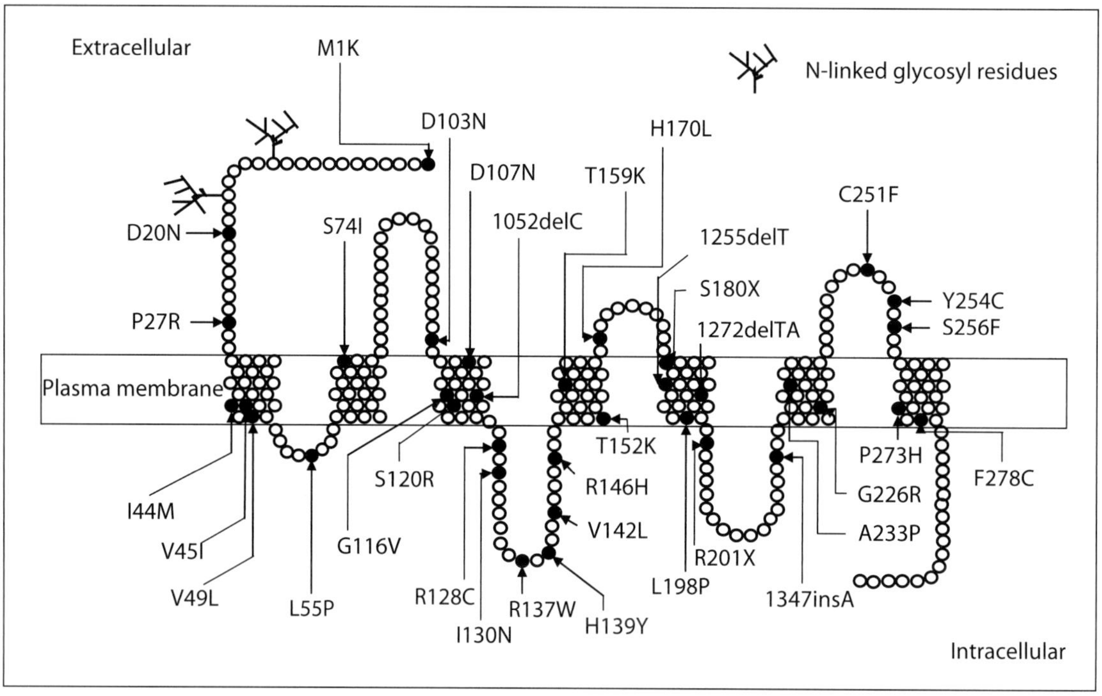

Fig. 3. Schematic 2-dimensional representation of human MC2R and mutations in FGD patients: the locations of all the polymorphisms and mutations published to date in patients with FGD are shown.

linkage of the disease to the MC2R locus on chromosome 18p11.2 [64, 65]. This implied that FGD was a genetically heterogeneous disease, and it was, therefore, categorized into two types: FGD type 1 with MC2R mutations and FGD type 2 with normal MC2R sequence [28]. Comparison of the clinical characteristics of FGD with MC2R mutations (FGD type 1) and without (FGD type 2) showed no significant differences either in the age of presentation of the patients or in the symptoms present at diagnosis. The ACTH and cortisol values were also comparable between the two subtypes [28].

The only unexpected difference was the height of these patients. Tall stature has been associated with FGD patients who have mutations in the MC2R. In contrast patients without MC2R mutations have heights that fall within the normal distribution [28]. The growth hormone (GH) and insulin-like growth factor (IGF-1) levels in the tall stature patients were normal [66]. The reasons for this tall stature are unclear but it has been suggested that the ACTH resistance that results from a defective MC2R may be associated with abnormalities of cartilage and/or bone growth independent of the GH-IGF-1 axis. It has also been implied that these abnormalities may be dependent on the actions of ACTH through other MCRs [67–69].

Melanocortin-2 Receptor Accessory Protein

In order to search for a gene associated with FGD type 2, a whole genome scan was carried out by SNP mapping using a 10K array [70]. Genomic DNA from the parents, unaffected and affected siblings of a family with FGD2 were genotyped in this study. Data analysis revealed a single candidate region at 21q22.1. Microsatellite marker analysis confirmed the homozygosity. RT-PCRs utilizing intron-skipping primers were carried out for all 30 known or predicted genes localized in this region using cDNA from human adrenal, brain and liver to seek adrenal-specific genes. Most genes were expressed in more than one tissue but a single gene was expressed only in the adrenal [70]. It encoded a novel single transmembrane domain protein of unknown function and was subsequently named the melanocortin-2 receptor accessory protein (MRAP).

The genomic sequence of MRAP consists of 6 exons of which the 5th or 6th exon are alternatively spliced to form MRAP-α or β, respectively. In humans MRAP-α encodes a protein of 172 amino acids and a predicted molecular weight of 19 kDa, and MRAP-β encodes a protein of 102 amino acids and a predicted molecular weight of 11.5 kDa. These isoforms are identical at the N-terminus and predicted transmembrane domain between residues 38 and 58, but thereafter are highly divergent.

MRAP Mutations in FGD Type 2 Patients

The entire exon sequence and the relevant flanking intronic regions were screened by DNA sequencing in 100 FGD patients without MC2R mutations [70]. Seven mutations were discovered in 21 families involving 25 patients (fig. 4). A missense mutation in the initiating methionine (M1I) was the most frequently occurring mutation. This was discovered in 9 patients from 8 families. Most of the other mutations were splice site mutations at the junction of exon3/intron3 (IVS3 + G>T, IVS3 + 1G>C, IVS3 + 1G>A, IVS3 + 1delG and IVS3 + 3insT). These account for 15 further patients from 12 families. The final mutation, a nonsense mutation (V44X) in exon 2, has only been found in 1 individual. Therefore MRAP mutations account for 25% of FGD type 2 cases and 15–20% of all FGD cases [70]. Several other mutations have since been identified (Y11X) in two siblings (our unpubl. data). All these mutations would result in either the complete absence of the protein or a severely truncated product.

Function of MRAP

Not much was known about the functional role of MRAP. It was originally identified as a protein named FALP that was upregulated on differentiation of mouse 3T3-L1 cells from preadipocytes into adipocytes [71]. Although the authors concluded that the function of the protein was not clear, they suggested that it might be involved in

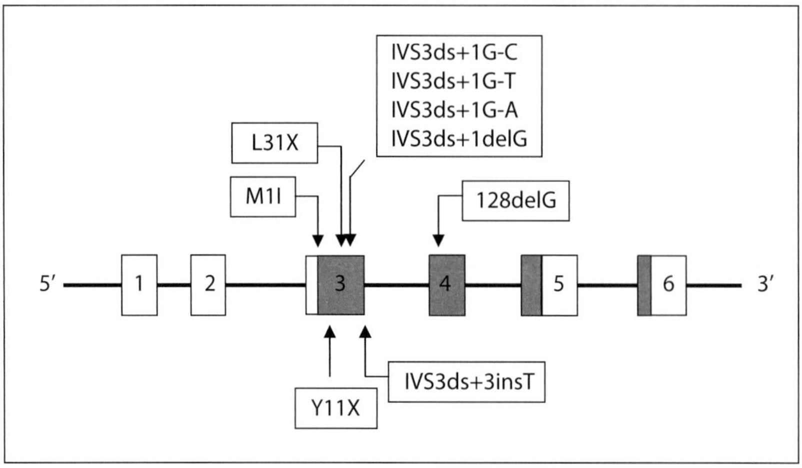

Fig. 4. Mutations in MRAP in patients with FGD type 2: gene structure indicating the location of mutations identified in 26 affected individuals from 21 families. Grey boxes indicate the location of the single open reading frame. Exons 5 and 6 are alternatively spliced to generate the MRAP-α and MRAP-β isoforms, respectively.

intracellular trafficking pathways. It is interesting to note that MC2R is also upregulated during the differentiation of the mouse 3T3-L1 cell line [72, 73].

MRAP appeared to be a good candidate as an accessory factor or chaperone for MC2R for several reasons. First, its expression pattern overlapped with that of MC2R with both proteins being expressed in the same tissues. Second, FGD patients with mutations in MC2R or MRAP show very similar clinical presentations [28]. Third, MRAP exhibited structural similarities to other GPCR accessory proteins such as receptor activity-modifying protein (RAMPs) and receptor transporter protein (RTPs), being small proteins with a single transmembrane domain.

MRAP Is Essential for MC2R Function

Confocal immunocytochemistry demonstrated that transfection of epitope-tagged MRAP and MC2R into non-adrenal cells such as CHO and SKN-SH cells resulted in the colocalization of the two proteins (fig. 5). MRAP facilitated the cell surface trafficking of MC2R, which would otherwise be retained at the endoplasmic reticulum [70]. The function of the MC2R in terms of cAMP production was assessed in cells transfected with the receptor in the presence and absence of MRAP. A significant increase in the MC2R-mediated cAMP response to ACTH was found in cells cotransfected with the two proteins (fig. 5). Co-immunoprecipitation studies using epitope-tagged MRAP and MC2R also showed that the two proteins could form homodimers [70, 74, 75].

The role of MRAP was ascertained further by RNAi studies using mouse Y1 adrenocortical cells, which express MRAP and a functional MC2R endogenously.

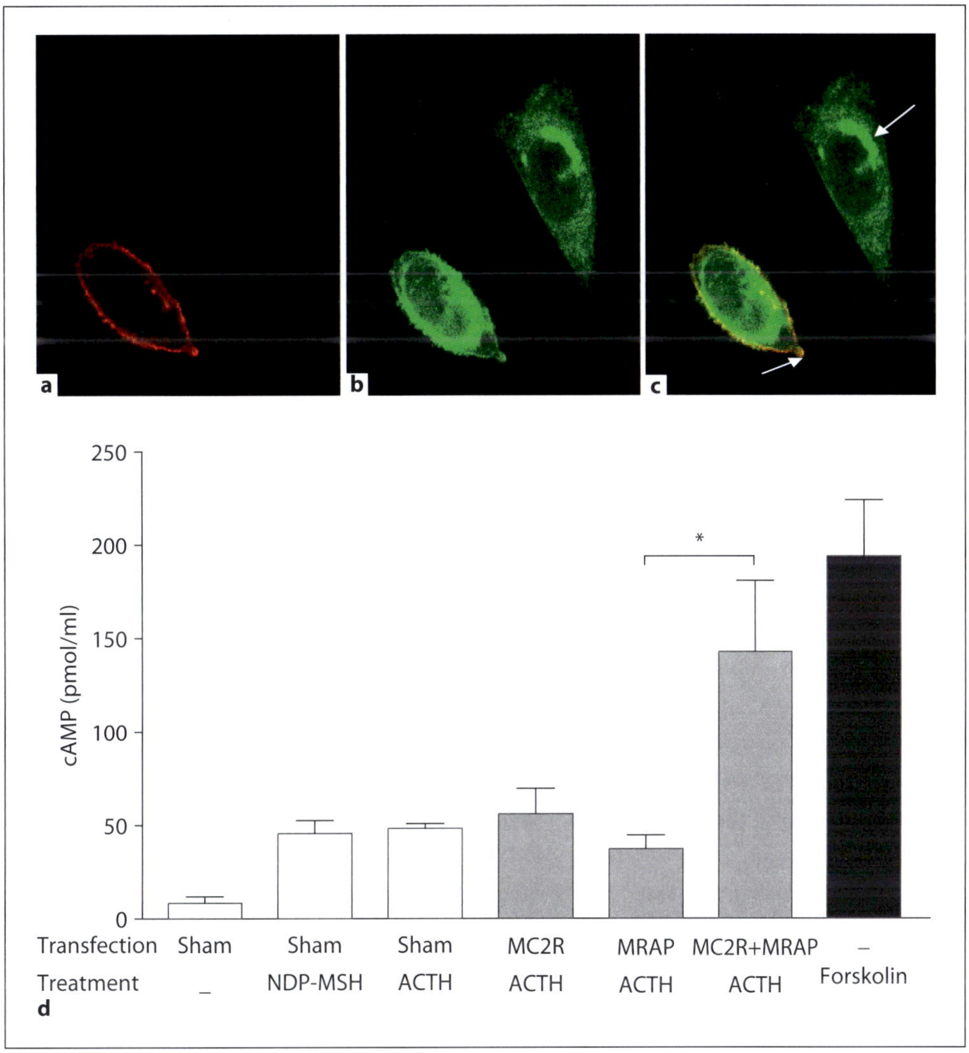

Fig. 5. MRAP in the functional expression of the MC2R: MC2R colocalizes with MRAP at the cell surface in the presence of MRAP. **a** Localization of MRAP-Flag (red). **b** Localization of MC2R-GFP (green). **c** Colocalization of the MC2R-GFP and MRAP-Flag in the merged image (yellow). **d** Functional cAMP assay: cells were either sham transfected with pcDNA 3.1 (white bars) or transfected with MC2R or MRAP alone or cotransfected with both MC2R-GFP and MRAP-Flag (grey bars). Forskolin was used as a positive control (black bar). Cells were left untreated or treated with 10^{-6} M ACTH, 10^{-6} M NDP-MSH or 10^{-5} M forskolin. Bars show the mean ± SEM of 3 individual experiments performed in duplicate. *$p < 0.05$.

MRAP siRNAs designed to target the first coding exon of MRAP were employed to reduce the expression of MRAP. Measurement of cellular cAMP levels showed that the MC2R function was impaired in cells that had the MRAP expression knocked down by the siRNA. This was further confirmed by creating stable cell lines that expressed MRAP shRNAs [75] (fig. 6). The reintroduction of the human form of

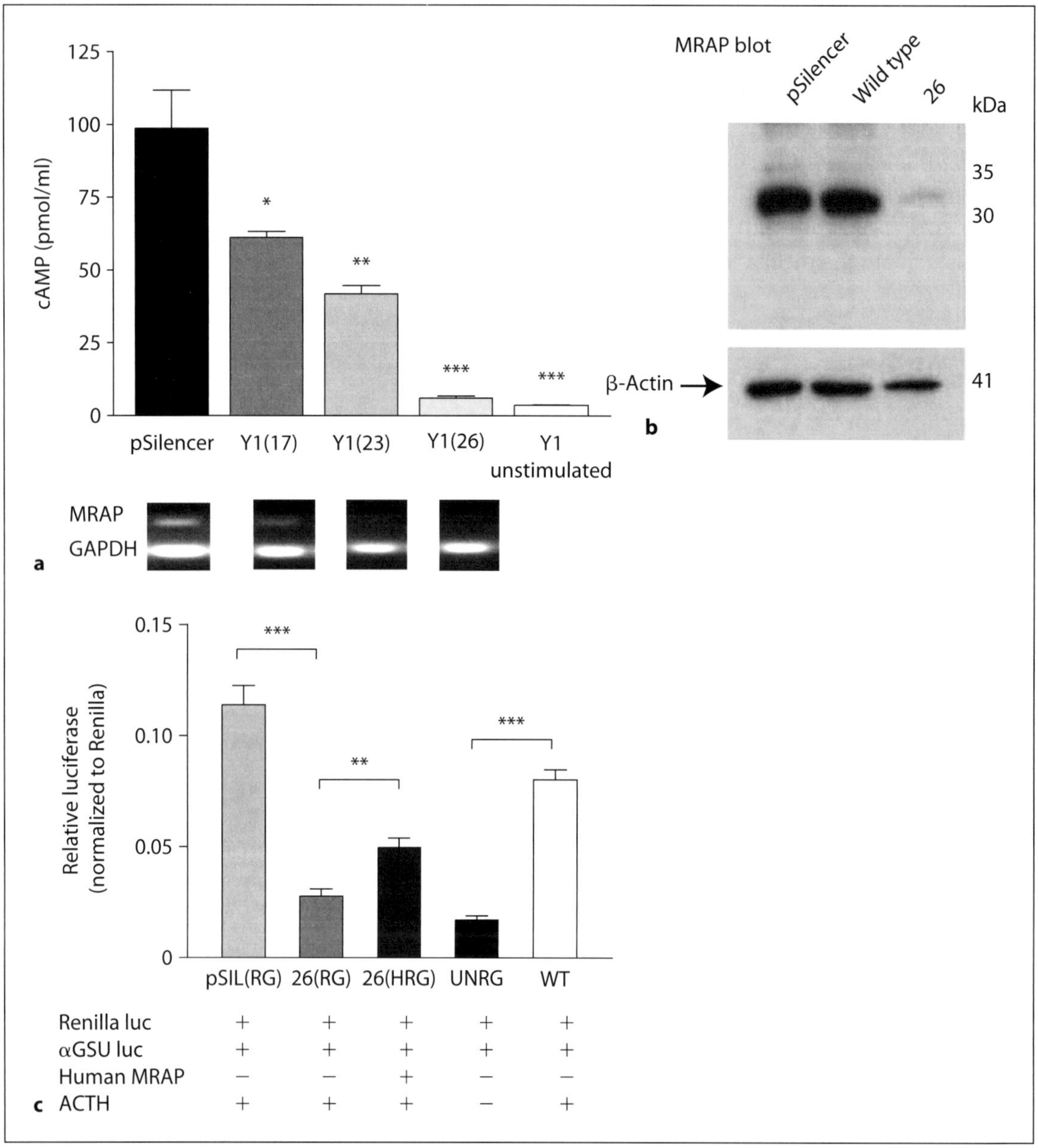

Fig. 6. Stable knockdown of MRAP using MRAP shRNA. **a** cAMP response in clonal cell lines expressing MRAP shRNAs. Stimulations were carried out in the presence of IBMX (10^{-5} M) with 10^{-6} M ACTH for 30 min. Y1 WT unstimulated cells were used as a control. Results are the mean of 3 individual experiments performed in duplicate. *$p < 0.05$, **$p < 0.01$, ***$p < 0.001$ compared to negative control (pSilencer). Shown in the panel below is the expression of MRAP on each of the clonal cell lines using RT-PCR. GAPDH was used as an internal control. **b** Immunoblot to show the knockdown of MRAP protein expression. Whole cell lysates from WT Y1 cells, WT cells transfected with pSilencer negative control vector and Y1 knockdown clone 26 cells were blotted with anti-MRAP antibody. β-Actin was used as a negative control. **c** Luciferase assay to assess the effect of MRAP expression on MC2R signalling in response to ACTH (10^{-6} M) in the clonal cell line 26 transfected with the human MRAP sequence. **$p < 0.01$; ***$p < 0.001$.

MRAP, which is resistant to silencing by mouse MRAP shRNAs, resulted in the restoration of MC2R signalling as determined by an increase in cAMP production (fig. 6). The means by which MRAP functions as an accessory protein for MC2R are currently under investigation but the data suggest that it functions as a unique antiparallel homodimer in a complex with the MC2R [74, 75]. Identification of other unidentified genes responsible for cases of FGD that have normal MC2R and MRAP may reveal further proteins that form part of a complex with MC2R and MRAP and help to elucidate the mechanism by which MRAP acts as an accessory protein.

Conclusion

ACTH resistance syndromes comprise a group of rare autosomal recessive and genetically heterogenic disorders. To date one gene responsible for the triple A syndrome (AAAS) and two genes responsible for ~45% FGD (MC2R and MRAP) have been identified but for both diseases further genes are implicated. Identification of these other genes will give a more detailed insight into the disease process leading to these syndromes. The discovery that MRAP is essential as an accessory protein for the function of MC2R not only provides interesting information into the processing, trafficking and signalling of this receptor but also extends our understanding of the biological aspects of GPCRs as a whole.

References

1 Li CH, Geschwind II, Levy AL, Harris JI, Dixon JS, Pon NG, Porath JO: Isolation and properties of alpha-corticotrophin from sheep pituitary glands. Nature 1954;173:251–253.

2 Lee TH, Lerner AB: Isolation of melanocyte-stimulating hormone from hog pituitary gland. J Biol Chem 1956;221:943–959.

3 Nakanishi S, Inoue A, Kita T, Nakamura M, Chang AC, Cohen SN, Numa S: Nucleotide sequence of cloned cDNA for bovine corticotropin-beta-lipotropin precursor. Nature 1979;278:423–427.

4 Raffin-Sanson ML, de Keyzer Y, Bertagna X: Proopiomelanocortin, a polypeptide precursor with multiple functions: from physiology to pathological conditions. Eur J Endocrinol 2003;149:79–90.

5 Chhajlani V, Wikberg JE: Molecular cloning and expression of the human melanocyte stimulating hormone receptor cDNA. FEBS Lett 1992;309:417–420.

6 Mountjoy KG, Robbins LS, Mortrud MT, Cone RD: The cloning of a family of genes that encode the melanocortin receptors. Science 1992;257:1248–1251.

7 Catania A, Gatti S, Colombo G, Lipton JM: Targeting melanocortin receptors as a novel strategy to control inflammation. Pharmacol Rev 2004;56:1–29.

8 Getting SJ: Targeting melanocortin receptors as potential novel therapeutics. Pharmacol Ther 2006; 111:1–15.

9 Rana BK: New insights into G-protein-coupled receptor signaling from the melanocortin receptor system. Mol Pharmacol 2003;64:1–4.

10 Wikberg JE, Muceniece R, Mandrika I, Prusis P, Lindblom J, Post C, Skottner A: New aspects on the melanocortins and their receptors. Pharmacol Res 2000;42:393–420.

11 Abdel-Malek ZA: Melanocortin receptors: their functions and regulation by physiological agonists and antagonists. Cell Mol Life Sci 2001;58:434–441.

12 Haynes RC Jr: The activation of adrenal phosphorylase by the adrenocorticotropic hormone. J Biol Chem 1958;233:1220–1222.

13 Lefkowitz RJ, Roth J, Pricer W, Pastan I: ACTH receptors in the adrenal: specific binding of ACTH-125I and its relation to adenyl cyclase. Proc Natl Acad Sci USA 1970;65:745–752.

14 Cammas FM, Kapas S, Barker S, Clark AJ: Cloning, characterization and expression of a functional mouse ACTH receptor. Biochem Biophys Res Commun 1995;212:912–918.

15 Xia Y, Wikberg JE: Localization of ACTH receptor mRNA by in situ hybridization in mouse adrenal gland. Cell Tissue Res 1996;286:63–68.

16 Hofmann K, Stehle CJ, Finn FM: Identification of a protein in adrenal particulates that binds adrenocorticotropin specifically and with high affinity. Endocrinology 1988;123:1355–1363.

17 Weber A, Kapas S, Hinson J, Grant DB, Grossman A, Clark AJ: Functional characterization of the cloned human ACTH receptor: impaired responsiveness of a mutant receptor in familial glucocorticoid deficiency. Biochem Biophys Res Commun 1993;197: 172–178.

18 Naville D, Barjhoux L, Jaillard C, Faury D, Despert F, Esteva B, Durand P, Saez J, Begeot M: Mutations of ACTH receptor gene and familial syndrome of glucocorticoid deficiency (in French). Ann Endocrinol (Paris) 1996;57:101–106.

19 Schioth HB, Chhajlani V, Muceniece R, Klusa V, Wikberg JE: Major pharmacological distinction of the ACTH receptor from other melanocortin receptors. Life Sci 1996;59:797–801.

20 Elias LL, Huebner A, Pullinger GD, Mirtella A, Clark AJ: Functional characterization of naturally occurring mutations of the human adrenocorticotropin receptor: poor correlation of phenotype and genotype. J Clin Endocrinol Metab 1999;84:2766–2770.

21 Yang YK, Ollmann MM, Wilson BD, Dickinson C, Yamada T, Barsh GS, Gantz I: Effects of recombinant agouti-signaling protein on melanocortin action. Mol Endocrinol 1997;11:274–280.

22 Kimura T: Effects of hypophysectomy and ACTH administration on the level of adrenal cholesterol side-chain desmolase. Endocrinology 1969;85: 492–499.

23 Waterman MR: Biochemical diversity of cAMP-dependent transcription of steroid hydroxylase genes in the adrenal cortex. J Biol Chem 1994;269: 27783–27786.

24 Waterman MR, Bischof LJ: Cytochromes P450 12: diversity of ACTH (CAMP)-dependent transcription of bovine steroid hydroxylase genes. FASEB J 1997;11:419–427.

25 Sewer MB, Waterman MR: Insights into the transcriptional regulation of steroidogenic enzymes and StAR. Rev Endocr Metab Disord 2001;2:269–274.

26 Kraemer FB: Adrenal cholesterol utilization. Mol Cell Endocrinol 2007;265–266:42–45.

27 Allgrove J, Clayden GS, Grant DB, Macaulay JC: Familial glucocorticoid deficiency with achalasia of the cardia and deficient tear production. Lancet 1978;i:1284–1286.

28 Clark AJ, Weber A: Adrenocorticotropin insensitivity syndromes. Endocr Rev 1998;19:828–843.

29 Metherell LA, Chan LF, Clark AJ: The genetics of ACTH resistance syndromes. Best Pract Res Clin Endocrinol Metab 2006;20:547–560.

30 Heinrichs C, Tsigos C, Deschepper J, Drews R, Collu R, Dugardeyn C, Goyens P, Ghanem GE, Bosson D, Chrousos GP, et al: Familial adrenocorticotropin unresponsiveness associated with alacrima and achalasia: biochemical and molecular studies in two siblings with clinical heterogeneity. Eur J Pediatr 1995; 154:191–196.

31 Tsigos C, Arai K, Latronico AC, DiGeorge AM, Rapaport R, Chrousos GP: A novel mutation of the adrenocorticotropin receptor (ACTH-R) gene in a family with the syndrome of isolated glucocorticoid deficiency, but no ACTH-R abnormalities in two families with the triple A syndrome. J Clin Endocrinol Metab 1995;80:2186–2189.

32 Weber A, Wienker TF, Jung M, Easton D, Dean HJ, Heinrichs C, Reis A, Clark AJ: Linkage of the gene for the triple a syndrome to chromosome 12q13 near the type II keratin gene cluster. Hum Mol Genet 1996;5:2061–2066.

33 Huebner A, Yoon SJ, Ozkinay F, Hilscher C, Lee H, Clark AJ, Handschug K: Triple A syndrome – clinical aspects and molecular genetics. Endocr Res 2000;26:751–759.

34 Tullio-Pelet A, Salomon R, Hadj-Rabia S, Mugnier C, de Laet MH, Chaouachi B, Bakiri F, Brottier P, Cattolico L, Penet C, Begeot M, Naville D, Nicolino M, Chaussain JL, Weissenbach J, Munnich A, Lyonnet S: Mutant WD-repeat protein in triple-A syndrome. Nat Genet 2000;26:332–335.

35 Handschug K, Sperling S, Yoon SJ, Hennig S, Clark AJ, Huebner A: Triple A syndrome is caused by mutations in AAAS, a new WD-repeat protein gene. Hum Mol Genet 2001;10:283–290.

36 Garcia-Higuera I, Fenoglio J, Li Y, Lewis C, Panchenko MP, Reiner O, Smith TF, Neer EJ: Folding of proteins with WD-repeats: comparison of six members of the WD-repeat superfamily to the G protein beta subunit. Biochemistry 1996;35:13985–13994.

37 Smith TF, Gaitatzes C, Saxena K, Neer EJ: The WD repeat: a common architecture for diverse functions. Trends Biochem Sci 1999;24:181–185.

38 Cronshaw JM, Krutchinsky AN, Zhang W, Chait BT, Matunis MJ: Proteomic analysis of the mammalian nuclear pore complex. J Cell Biol 2002;158:915–927.

39 Allen TD, Cronshaw JM, Bagley S, Kiseleva E, Goldberg MW: The nuclear pore complex: mediator of translocation between nucleus and cytoplasm. J Cell Sci 2000;113:1651–1659.

40 Dreger M, Bengtsson L, Schoneberg T, Otto H, Hucho F: Nuclear envelope proteomics: novel integral membrane proteins of the inner nuclear membrane. Proc Natl Acad Sci USA 2001;98:11943–11948.

41 Cronshaw JM, Matunis MJ: The nuclear pore complex: disease associations and functional correlations. Trends Endocrinol Metab 2004;15:34–39.

42 Cronshaw JM, Matunis MJ: The nuclear pore complex protein ALADIN is mislocalized in triple A syndrome. Proc Natl Acad Sci USA 2003;100:5823–5827.

43 Huebner A, Kaindl AM, Knobeloch KP, Petzold H, Mann P, Koehler K: The triple A syndrome is due to mutations in ALADIN, a novel member of the nuclear pore complex. Endocr Res 2004;30:891–899.

44 Krumbholz M, Koehler K, Huebner A: Cellular localization of 17 natural mutant variants of ALADIN protein in triple A syndrome – shedding light on an unexpected splice mutation. Biochem Cell Biol 2006; 84:243–249.

45 Huebner A, Mann P, Rohde E, Kaindl AM, Witt M, Verkade P, Jakubiczka S, Menschikowski M, Stoltenburg-Didinger G, Koehler K: Mice lacking the nuclear pore complex protein ALADIN show female infertility but fail to develop a phenotype resembling human triple A syndrome. Mol Cell Biol 2006;26: 1879–1887.

46 Hirano M, Furiya Y, Asai H, Yasui A, Ueno S: ALADINI482S causes selective failure of nuclear protein import and hypersensitivity to oxidative stress in triple A syndrome. Proc Natl Acad Sci USA 2006; 103:2298–2303.

47 Grant DB, Barnes ND, Dumic M, Ginalska-Malinowska M, Milla PJ, von Petrykowski W, Rowlatt RJ, Steendijk R, Wales JHK, Werder E: Neurological and adrenal dysfunction in the adrenal insufficiency/alacrima/achalasia (3A) syndrome. Arch Dis Childhood 1993;68:779–782.

48 Shepard TH, Landing BH, Mason DG: Familial Addison's disease; case reports of two sisters with corticoid deficiency unassociated with hypoaldosteronism. AMA J Dis Child 1959;97:154–162.

49 Kelch RP, Kaplan SL, Biglieri EG, Daniels GH, Epstein CJ, Grumbach MM: Hereditary adrenocortical unresponsiveness to adrenocorticotropic hormone. J Pediatr 1972;81:726–736.

50 Migeon CJ, Kenny EM, Kowarski A, Snipes CA, Spaulding JS, Finkelstein JW, Blizzard RM: The syndrome of congenital adrenocortical unresponsiveness to ACTH. Report of six cases. Pediatr Res 1968; 2:501–513.

51 Moshang T Jr, Rosenfield RL, Bongiovanni AM, Parks JS, Amrhein JA: Familial glucocorticoid insufficiency. J Pediatr 1973;82:821–826.

52 Franks RC, Nance WE: Hereditary adrenocortical unresponsiveness to ACTH. Pediatrics 1970;45: 43–48.

53 Lacy DE, Nathavitharana KA, Tarlow MJ: Neonatal hepatitis and congenital insensitivity to adrenocorticotropin (ACTH). J Pediatr Gastroenterol Nutr 1993;17:438–440.

54 Lin L, Hindmarsh PC, Metherell LA, Alzyoud M, Al Ali M, Brain CE, Clark AJ, Dattani MT, Achermann JC: Severe loss-of-function mutations in the adrenocorticotropin receptor (ACTHR, MC2R) can be found in patients diagnosed with salt-losing adrenal hypoplasia. Clin Endocrinol (Oxf) 2007;66: 205–210.

55 Xia Y, Wikberg JE: Localization of ACTH receptor MRNA by in situ hybridization in mouse adrenal gland. Cell Tissue Res 1996;286:63–68.

56 Arvat E, Di Vito L, Lanfranco F, Maccario M, Baffoni C, Rossetto R, Aimaretti G, Camanni F, Ghigo E: Stimulatory effect of adrenocorticotropin on cortisol, aldosterone, and dehydroepiandrosterone secretion in normal humans: dose-response study. J Clin Endocrinol Metab 2000;85:3141–3146.

57 Chida D, Nakagawa S, Nagai S, Sagara H, Katsumata H, Imaki T, Suzuki H, Mitani F, Ogishima T, Shimizu C, Kotaki H, Kakuta S, Sudo K, Koike T, Kubo M, Iwakura Y: Melanocortin 2 receptor is required for adrenal gland development, steroidogenesis, and neonatal gluconeogenesis. Proc Natl Acad Sci USA 2007;104:18205–18210.

58 Clark AJ, McLoughlin L, Grossman A: Familial glucocorticoid deficiency associated with point mutation in the adrenocorticotropin receptor. Lancet 1993;341:461–462.

59 Weber A, Toppari J, Harvey RD, Klann RC, Shaw NJ, Ricker AT, Nanto-Salonen K, Bevan JS, Clark AJ: Adrenocorticotropin receptor gene mutations in familial glucocorticoid deficiency: relationships with clinical features in four families. J Clin Endocrinol Metab 1995;80:65–71.

60 Naville D, Penhoat A, Barjhoux L, Jaillard C, Fontanay S, Saez J, Durand P, Begeot M: Characterization of the human ACTH receptor gene and in vitro expression. Endocr Res 1996;22:337–348.

61 Tsigos C, Arai K, Hung W, Chrousos GP: Hereditary isolated glucocorticoid deficiency is associated with abnormalities of the adrenocorticotropin receptor gene. J Clin Invest 1993;92:2458–2461.

62 Slavotinek AM, Hurst JA, Dunger D, Wilkie AO: ACTH receptor mutation in a girl with familial glucocorticoid deficiency. Clin Genet 1998;53:57–62.

63 Fluck CE, Martens JW, Conte FA, Miller WL: Clinical, genetic, and functional characterization of adrenocorticotropin receptor mutations using a novel receptor assay. J Clin Endocrinol Metab 2002;87: 4318–4323.

64 Weber A, Clark AJ: Mutations of the ACTH receptor gene are only one cause of familial glucocorticoid deficiency. Hum Mol Genet 1994;3:585–588.

65 Naville D, Weber A, Genin E, Durand P, Clark AJ, Begeot M: Exclusion of the adrenocorticotropin (ACTH) receptor (MC2R) locus in some families with ACTH resistance but no mutations of the MC2R coding sequence (familial glucocorticoid deficiency type 2). J Clin Endocrinol Metab 1998; 83:3592–3596.
66 Elias LL, Huebner A, Metherell LA, Canas A, Warne GL, Bitti ML, Cianfarani S, Clayton PE, Savage MO, Clark AJ: Tall stature in familial glucocorticoid deficiency. Clin Endocrinol (Oxf) 2000;53:423–430.
67 Mountjoy KG, Wild JM: Melanocortin-4 receptor MRNA expression in the developing autonomic and central nervous systems. Brain Res Dev Brain Res 1998;107:309–314.
68 Imamine H, Mizuno H, Sugiyama Y, Ohro Y, Sugiura T, Togari H: Possible relationship between elevated plasma ACTH and tall stature in familial glucocorticoid deficiency. Tohoku J Exp Med 2005; 205:123–131.
69 Zhong Q, Sridhar S, Ruan L, Ding KH, Xie D, Insogna K, Kang B, Xu J, Bollag RJ, Isales CM: Multiple melanocortin receptors are expressed in bone cells. Bone 2005;36:820–831.
70 Metherell LA, Chapple JP, Cooray S, David A, Becker C, Ruschendorf F, Naville D, Begeot M, Khoo B, Nurnberg P, Huebner A, Cheetham ME, Clark AJ: Mutations in MRAP, encoding a new interacting partner of the ACTH receptor, cause familial glucocorticoid deficiency type 2. Nat Genet 2005;37:166–170.
71 Xu A, Choi KL, Wang Y, Permana PA, Xu LY, Bogardus C, Cooper GJ: Identification of novel putative membrane proteins selectively expressed during adipose conversion of 3T3-L1 cells. Biochem Biophys Res Commun 2002;293:1161–1167.
72 Boston BA, Cone RD: Characterization of melanocortin receptor subtype expression in murine adipose tissues and in the 3T3-L1 cell line. Endocrinology 1996;137:2043–2050.
73 Noon LA, Clark AJ, King PJ: A peroxisome proliferator-response element in the murine mc2-r promoter regulates its transcriptional activation during differentiation of 3T3-L1 adipocytes. J Biol Chem 2004;279:22803–22808.
74 Sebag JA, Hinkle PM: Melanocortin-2 receptor accessory protein MRAP forms antiparallel homodimers. Proc Natl Acad Sci USA 2007;104:20244–20249.
75 Cooray SN, Almiro Do Vale I, Leung K, Webb TR, Chapple JP, Egertova MR, Cheetham ME, Elphick M, Clark AJ: The melanocortin 2 receptor accessory protein exists as a homodimer and is essential for the function of the melanocortin 2 receptor in the mouse Y1 cell line. Endocrinology 2008;149:1935–1941.

Prof. Adrian J.L. Clark
Centre for Endocrinology, William Harvey Research Institute Barts and the London
Queen Mary University of London
Charterhouse Square
London EC1M 6BQ (UK)
Tel. +44 2078 826 202, Fax +44 278 826 197, E-Mail a.j.clark@qmul.ac.uk

Flück CE, Miller WL (eds): Disorders of the Human Adrenal Cortex.
Endocr Dev. Basel, Karger, 2008, vol 13, pp 117–132

Cushing Syndrome Caused by Adrenocortical Tumors and Hyperplasias (Corticotropin-Independent Cushing Syndrome)

Constantine A. Stratakis

Section on Endocrinology and Genetics, Program on Developmental Endocrinology and Genetics, National Institute of Child Health and Human Development, National Institutes of Health, Bethesda, Md., USA

Abstract

Endogenous Cushing syndrome (CS) is caused by excess adrenal glucocorticoid secretion that is adrenocorticotropin (ACTH)-dependent or independent; ACTH-independent adrenocortical causes of CS account for up to 20% of CS in adults, and 15% in children over age 7 years. In younger children, ACTH-independent CS may account for as many as half of the CS cases. In both adults and children, adrenocortical lesions causing CS include the common, isolated and sporadic, solitary cortisol-producing adenoma, the rare adrenocortical cancer, and a spectrum of recently recognized, bilateral hyperplasias (bilateral adrenocortical hyperplasias, BAHs): micronodular adrenal disease and its pigmented variant, primary pigmented nodular adrenocortical disease are mostly genetic processes. Macronodular BAHs, ACTH-independent macronodular hyperplasia or massive macronodular adrenocortical disease are less frequently genetic and almost never present in children (except in McCune-Albright syndrome); they present often with atypical CS in middle-aged or elderly adults. The majority of benign adrenocortical tumors associated with CS are associated with defects of the cAMP signaling pathway, whereas adrenal cancer is linked to aberrant expression of growth factors and germline or somatic mutations of tumor suppressor genes such as *TP53*. Adrenalectomy is the preferred mode of treatment for all adrenocortical causes of CS.

Endogenous Cushing syndrome (CS) is a rare disorder [1]. There are significant differences in the pathophysiology and epidemiology of hypercortisolemia among age groups [2]; different criteria are being used for the confirmation and differential diagnosis of this disorder in children [3, 4]. CS may be caused by corticotropin (ACTH)-producing pituitary tumors, a disorder also known as 'Cushing disease', or by ACTH-independent, cortisol-producing adrenocortical tumors (ADTs). In children, ectopic production of ACTH is extraordinarily rare [2, 5]: it has been reported only in a handful of cases confined at the extremes of pediatric age, infants with neuroblastomas or other neuroendocrine tumors [5] and adolescents with carcinoids, sporadic or in the context of multiple endocrine neoplasia type 1. Bilateral adrenocortical hyperplasias (BAHs) are far more common as causes of CS in children than in older patients [1, 2].

Epidemiology

Adrenocortical neoplasms account for less than 0.5% of all clinically significant tumors; however, autopsy studies indicate that as many as 10% of adults over the age of 40 years may have an ADT, usually a simple nodule that is not larger than 1 cm; up to 36% may have micronodular hyperplasia [6]. CS is a manifestation of approximately one third of all ADTs. In children, a significant number of ADTs presenting with CS are malignant, but the opposite is true in adults. There is a female-to-male predominance for ADTs in all ages (although this is probably not true for infants and toddlers).

Clinical Presentation

In most patients the onset of CS is rather insidious [1–4]. The most common presenting symptom of the syndrome is weight gain, but it is not universally present. Pathognomonic for CS in childhood is weight gain associated with growth retardation [7]. Other common problems reported include facial plethora, headaches, hypertension, hirsutism, amenorrhea, and hypogonadism (or delayed sexual maturation in children) [8]. Virilization is rare in ADTs unless the tumor produces adrenal androgens in addition to glucocorticoids; skin manifestations, including acne, violaceous striae and bruising and acanthosis nigricans, are also common. Sleep disruption, muscular weakness and mental changes are frequent.

Diagnostic Evaluation of ACTH-Independent CS

Diagnostic testing for CS is complicated by developmental differences in the regulation of hypothalamic-pituitary-adrenal axis between young and older individuals, as well as other factors, such as exposure to steroid hormones, medications and other exogenous substances, stress, and chronic illness [3, 8]; these factors may influence normal values for several of the tests employed in the workup for CS [2–4, 8]. In children, the diagnosis of CS is facilitated by the inhibitory effects of hypercortisolemia on height gain [7]. Indeed, the deceleration of growth velocity with a concurrent and unabated weight gain are the hallmark of CS in childhood [7–9]. In contrast, in lieu of these apparent signs, extensive biochemical investigation is needed for the confirmation of the diagnosis of CS in adult patients [1–4], especially if their situation is

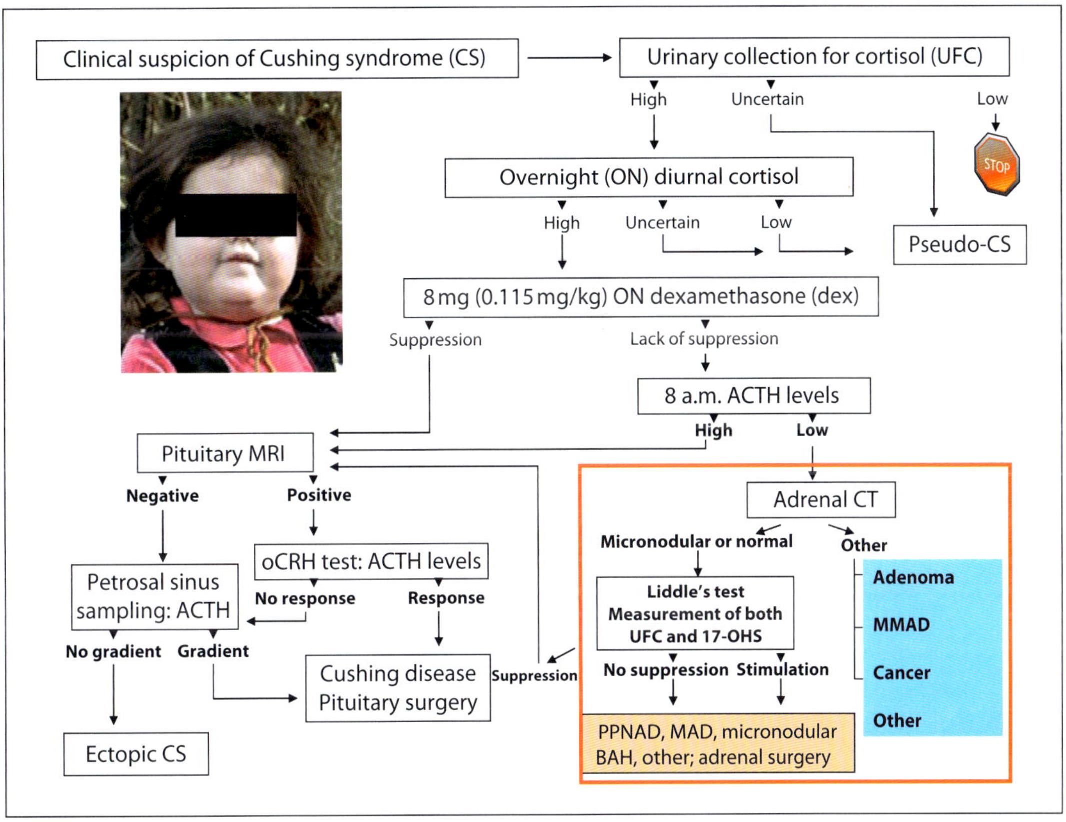

Fig. 1. Diagnostic workup of CS; in the box (right lower panel) adrenocortical causes of CS are listed.

complicated by moderate weight gain and other conditions, which collectively have been called 'pseudo-Cushing' states [10]. Adult patients may also have ectopic sources of ACTH as causes of CS in up to 10–15% of the total number of cases [11]. In general, the diagnostic evaluation proposed in figure 1 is what we recommend for both adults and children with CS [1–4, 8, 10–12] (fig. 1).

The appropriate therapeutic interventions in CS depend on accurate diagnosis and classification of the disease. The history and clinical evaluation (including growth charts in children) are important to make the initial diagnosis. Upon suspicion of the syndrome, laboratory and imaging confirmations are necessary. An algorithm of the diagnostic process is presented in figure 1. The first step in the diagnosis of CS is to document hypercortisolism. This step is usually done in the outpatient setting. Because of the circadian nature of cortisol and ACTH, isolated cortisol and ACTH measurements are not of great value in diagnosis. One excellent screening test for hypercortisolism is a 24-hour urinary free cortisol (UFC) excretion corrected for body surface area. A normal 24-hour UFC value is $<70\ \mu g/m^2/day$ (with the radioimmunoassay values). Falsely high UFC may be obtained because of physical and emotional stress, chronic and severe obesity, pregnancy, chronic exercise, depression, alcoholism, anorexia, narcotic withdrawal, anxiety, malnutrition and excessive water intake (more than 5 liters/day). These conditions may lead to sufficiently high UFCs to cause what is known as pseudo-CS. On the other hand, falsely low UFC may be obtained mostly with inadequate collection. Another baseline test for the establishment of the diagnosis of CS is a low-dose dexamethasone suppression test; the cortisol cutoff level should be $<1.8\,\mu g/dl$ (50 nmol/l); if it is greater than 1.8 μg/dl, further evaluation is necessary. If the response to both the 1 mg dexamethasone overnight suppression test and the 24-hour UFC are both normal, a diagnosis of CS may be excluded with the following caveat: 5–10% of patients may have intermittent or periodic cortisol hypersecretion and may not manifest abnormal results to either test. If periodic or intermittent CS is suspected, continuous follow-up of the patients is recommended. Diurnal plasma cortisol variation, including midnight cortisol values, is a very good test for the establishment of the diagnosis of CS: in our institution, it has become the test of choice for the confirmation of endogenous hypercortisolemia and is routinely done in patients with confirmed elevated urinary cortisol levels on the outside. There are several caveats for the interpretation of the test of which the most important ones are: (1) the venous catheter has to be placed at least 2 h before the test and (2) if the patient comes from another time zone, a 1-h/day adjustment should be taken into account prior to obtaining the test. In general, serum cortisol levels are drawn at 11:30 p.m. and 12:00 midnight and at 7:30 a.m. and 8:00 a.m., while the patient is lying in bed and asleep; midnight cortisol levels above 5 μg/dl are abnormal and confirm the diagnosis of CS, whereas an inverted diurnal rhythm is seen in BAHs and some other adrenal tumors. If one of the tests suggests CS or if there is any question about the diagnosis, tests that distinguish between pseudo-Cushing states and CS may be obtained. One such test is the combined dexamethasone-CRH test. Once the diagnosis of CS is confirmed there are several tests to distinguish ACTH-dependent disease from the ACTH-independent syndrome. A spot plasma ACTH may be measured; if this measurement is <5 pmol/l it is indicative of ACTH-independent CS, although the sensitivity and specificity of a single ACTH measurement are not high because of the great variability in plasma ACTH levels and the instability of the molecule after the sample's collection. Even if one assumes that the sample was collected and processed properly (collected on ice and spun down immediately in a refrigerated centrifuge for plasma separation; the sample should then be immediately processed or frozen at $-20°C$), ACTH levels that are between 5 and 20 pmol/l are not informative in this era of high sensitivity assays; levels above 20 pmol/l are more suggestive of an ACTH-dependent condition, but again that is not a certainty until single ACTH levels are repeatedly over 70 pmol/l.

The standard 6-day low- and high-dose dexamethasone suppression test (Liddle's test) is used to differentiate Cushing disease from ectopic ACTH secretion and adrenal causes of CS. In the classic form of this test, after 2 days of baseline urine collection, 0.5 mg of dexamethasone (adjusted per weight for children <70 kg by dividing the dose by 70 and multiplying by the weight of the child) every 6 h are given per os starting at 6.00 a.m. on day 3 ('low-dose' phase of the test) for a total of 8 doses (2 days); this is continued with a 2-mg dose of dexamethasone per os (adjusted per weight for children <70 kg by dividing the dose by 70 and multiplying by the weight of the child) on day 5 ('high-dose' phase of the test) given every 6 h for another 8 doses (final 2 days). UFCs and 17-hydroxysteroid (17-OHS) excretion are measured at baseline, during, and 1 day after the end of the dexamethasone administration. Approximately 90% of patients with Cushing disease will have suppression of cortisol and 17-OHS values, whereas less than 10% of patients with ectopic ACTH secretion will have suppression. UFC values should suppress to 90% of baseline value and 17-OHS excretion should suppress to less than 69% of baseline value. The criteria are similar if one uses serum cortisol values obtained at 8 a.m. of the morning after the last dose of dexamethasone, e.g. serum cortisol on day 7 should be 90% of baseline serum cortisol values (obtained at 8 a.m. the day before dexamethasone administration).

The Liddle test has been modified to (1) giving 2 mg every 6 h (without the preceding low-dose phase); (2) administering dexamethasone intravenously over 5 h at a rate of 1 mg/h, or (3) giving a single high dose of dexamethasone (8 mg, in children adjusted for weight <70 kg) at 11 p.m. and measuring the plasma cortisol level the following morning. This overnight, high-dose dexamethasone test has sensitivity and specificity values similar to those of the classic Liddle test: a 68% suppression of serum cortisol levels from baseline is what differentiates Cushing disease (more than 68% suppression) from other causes of CS (adrenal or ectopic ACTH production; less than 50% suppression) [12]. An ovine CRH stimulation test may also be obtained for the differentiation of Cushing disease from ectopic ACTH secretion [11], but it is less useful in the diagnosis of ADTs [8].

In addition to the biochemical testing, the most useful tests in the diagnosis of cortisol-producing ADTs are imaging [computed tomography (CT) or magnetic resonance imaging (MRI)] and Liddle's test (especially in the diagnosis of BAHs). CT is preferred over MRI for cortical (vs. medullary) tumors because it allows for better delineation of the adrenal contour (fig. 2). With the use of contrast material, adrenal CT is an excellent diagnostic tool in the investigation of ADTs although more testing is required for the delineation of a cortisol-producing benign ADT. Most adrenocortical carcinomas are unilateral and quite large by the time they are detected. Ultrasound may not be used to image the adrenal glands for the diagnostic workup of CS, because its sensitivity and accuracy are much less than CT or MRI. Catheterization studies may not be used to confirm the source of cortisol secretion in ADTs.

Histological Types of Benign ADTs Causing CS

Benign ADTs causing CS include the common adrenocortical adenoma (ADA) and BAHs [13] such as primary pigmented nodular adrenocortical disease (PPNAD) and

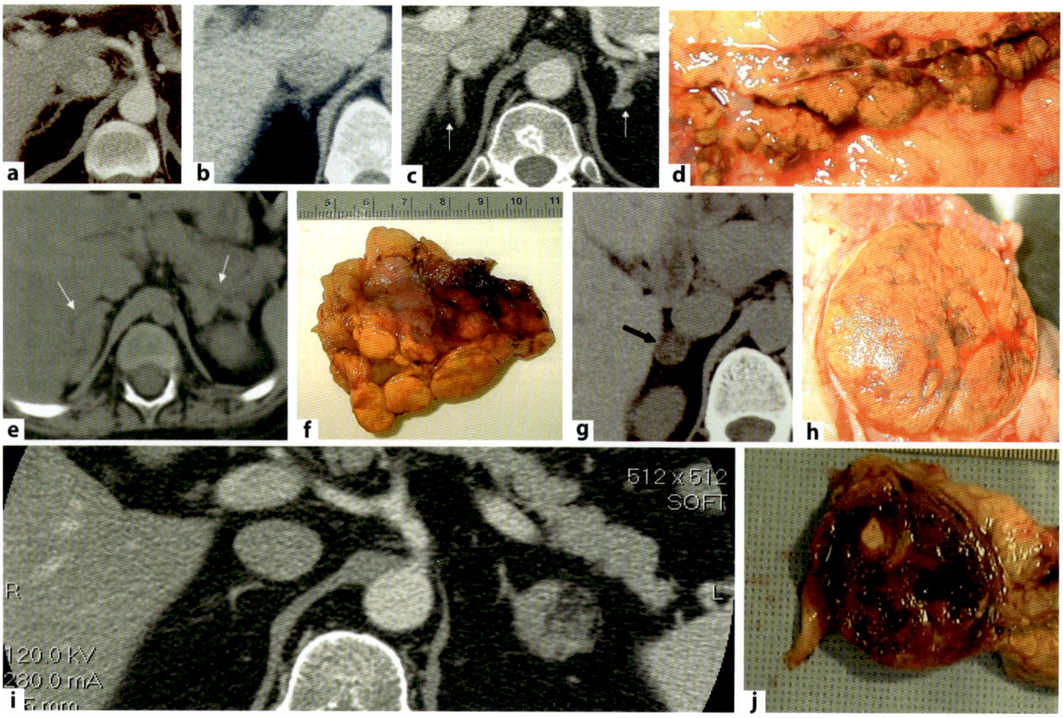

Fig. 2. a–c CT of the adrenal glands from 3 patients with micronodular BAHs. **d** The right adrenal gland with PPNAD from the patient whose CT is shown in **c**: multiple small, pigmented nodules are seen that are characteristic of PPNAD. CT from a patient with MMAD/AIMAH and CS with massive enlargement of the adrenal glands on both sides (arrows) (**e**) and her left adrenal with visible multiple macronodules (**f**). **g** CT from an adolescent with a common, solitary, cortisol-producing adrenal adenoma (arrow). **h** The tumor from the patient in **g**: this is a classic cortisol-producing adenoma with yellow fat and brown discoloration, the result of pigment (lipofuscin and rarely neuromelanin). **i** CT from a patient with a left-sided cortisol-producing tumor that was inhomogeneous with hemorrhage and calcifications. **j** The tumor from the patient in **i**: calcification and blood are visible; the tumor was premalignant.

ACTH-independent macronodular adrenocortical hyperplasia (AIMAH) – also known as massive macronodular adrenocortical disease (MMAD). The various types of adrenocortical lesions, their histology and other information are given in table 1. The common cortisol-producing ADA of the zona fasciculata needs little introduction, although histological variants of this common lesion do exist.

In all ages, the most common ADT causing CS is a unilateral adenoma (fig. 2); however, up to 10% of patients may have bilateral tumors [13]. Table 1 lists no less than 6 types of BAHs. They are divided into two groups of disorders, macro- and micronodular hyperplasias on the basis of the size of the associated nodules (fig. 2). In macronodular disorders, the greatest diameter of each nodule exceeds 1 cm; in the

Table 1. Adrenocortical causes of CS

Adrenocortical lesions	Age group	Histopathology	Genetics	Gene/locus
Benign				
Common adenoma	all ages	adenoma of the zona fasciculata	MEN 1, FAP, MAS, HLRCS, CNC, Carney triad, other	menin, *APC*, *GNAS*, FH, *PRKAR1A*; 2p16, 9q34, other
Macronodular hyperplasias (multiple nodules more than 1 cm each)				
Bilateral macroadenomatous hyperplasia (BMAH)	middle age	distinct adenomas (usually 2 or 3) with internodular atrophy	MEN 1, FAP, MAS, HLRCS, other; isolated (AD); other	menin, *APC*, *GNAS*, FH, ectopic GPCRs
BMAH of childhood (c-BMAH)	infants, very young children	as above; occasional microadenomas	MAS	*GNAS*
AIMAH, also known as MMAD (AIMAH/MMAD)	middle age	adenomatous hyperplasia (multiple) with internodular hyperplasia of the zona fasciculata	isolated, AD	ectopic GPCRs; WISP-2 and Wnt signaling; 17q22-24, other
Micronodular hyperplasias (multiple nodules less than 1 cm each)				
Isolated PPNAD (i-PPNAD)	children; young adults	microadenomatous hyperplasia with (mostly) internodular atrophy and nodular pigment (lipofuscin)	isolated; AD	*PRKAR1A*, *PDE11A*; 2p16; other
CNC-associated primary pigmented nodular adrenocortical disease (c-PPNAD)	children; young and middle ages	microadenomatous hyperplasia with (mostly) internodular atrophy and (mainly nodular) pigment (lipofuscin)	CNC (AD)	*PRKAR1A*, 2p16; other
Isolated micronodular adrenocortical disease (i-MAD)	mostly children; young adults	microadenomatous with hyperplasia of the surrounding zona fasciculata and limited or absent pigment	isolated, AD; other	*PDE11A*, *PDE8B*, other; 2p12-p16, 5q, other

Table 1. (continued)

Adrenocortical lesions	Age group	Histopathology	Genetics	Gene/locus
Malignant				
Cancer (sporadic)	all ages	mitotic figures, atypia of cortical cells; capsular invasion; metastases	isolated	*TP53*, β-catenin, *INHA*; 2p, 2q, 9q, 11q, other
Cancer (syndromic)	children; young adults	as above	LFS (AD); BWS, RTS, other	*TP53, CHK22, IGF2*, other
Brazil variant	children; young adults	as above; milder clinical course	AD; other	*TP53, INHA*, SF1; 9q34 amplification; other

MEN 1 = Multiple endocrine neoplasia type 1; FAP = familial adenomatous polyposis (polyposis coli); MAS = McCune-Albright syndrome; HLRCS = hereditary leiomyomatosis and renal cancer syndrome; FH = fumarate hydratase; AD = autosomal dominant; CNC = Carney complex; GPCR = G-protein-coupled receptors; LFS = Li-Fraumeni syndrome; BWS = Beckwith-Wiedemann syndrome; RTS = Rubinstein-Taybi syndrome.

micronodular group nodules are less than 1 cm. Although nodules less than 1 cm can occur in macronodular disease (especially the form associated with McCune-Albright syndrome), and single large tumors may be encountered in PPNAD (especially in older patients), the size criterion has biologic relevance, as we rarely see a continuum in the same subject: most patients are either macro- or micronodular. There are two additional basic characteristics that we use in this classification of BAHs [13]: that of the presence of pigment and that of status (hyperplasia or atrophy) of the surrounding cortex. Pigment in adrenocortical lesions is rarely melanin; most of the pigmentation in both ADAs and BAH that produce cortisol is lipofuscin (fig. 2). The latter appears macroscopically as light brown to, some times, dark brown or even black discoloration of the tumorous or hyperplastic tissue; microscopically, lipofuscin can be seen but it is better detected by electron microscopy.

PPNAD is a genetic disorder with the majority of cases associated with Carney complex, a syndrome of multiple endocrine gland abnormalities in addition to lentigines and myxomas; the adrenal glands in PPNAD are most commonly normal or even small in size with multiple pigmented nodules surrounded by an atrophic cortex (fig. 3). The nodules are autonomously functioning resulting in the surrounding atrophy of

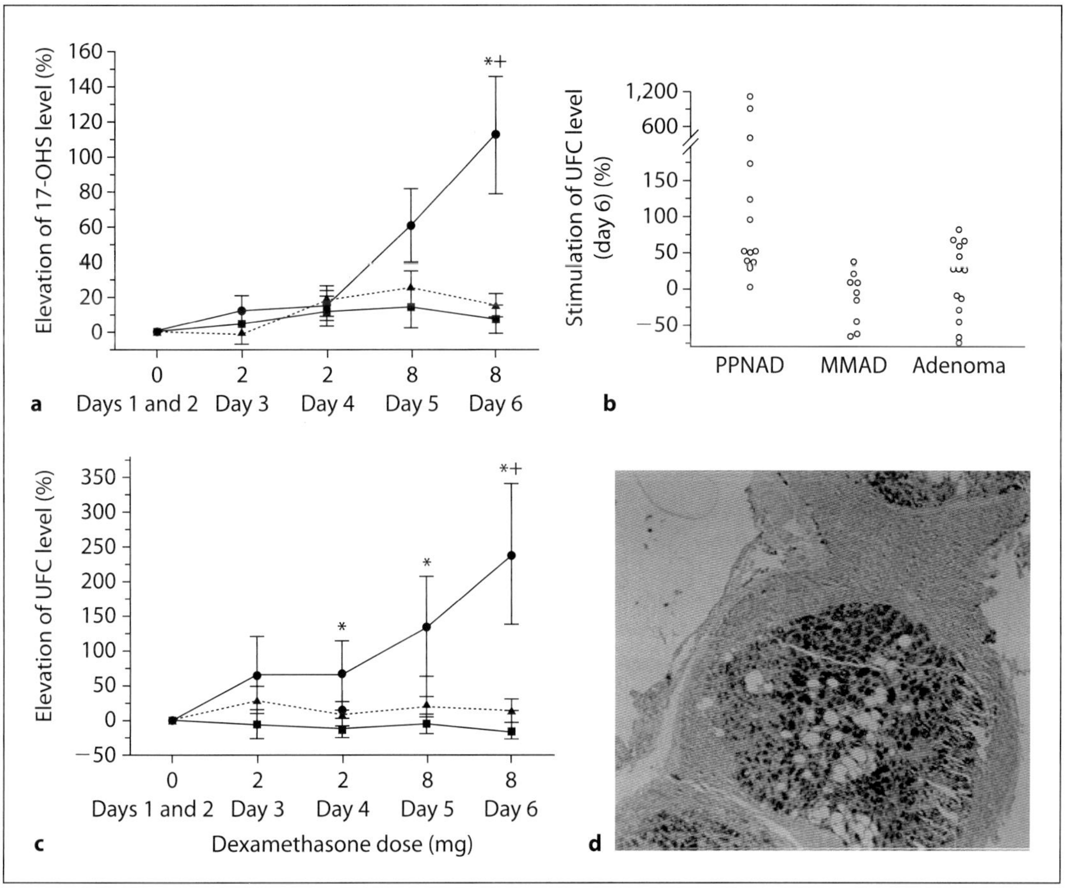

Fig. 3. Paradoxical stimulation of 17-OHS (**a**) and UFC (**c**) in patients with PPNAD (l) versus those with MMAD (d) and solitary cortisol-producing adenomas (+) during the course of Liddle's test. *p > 0.05; +p > 0.001. **b** 50% increase of UFCs on day 6 of the test detects more than 70% of the patients with PPNAD versus other cortisol-producing tumors of the adrenal cortex. **d** PPNADs express synaptophysin within the cortical nodules, an unexpected feature for a cortical lesion.

the cortex. Children and adolescents with PPNAD frequently have periodic or atypical CS [14, 15].

AIMAH/MMAD is another rare disease, which leads to CS [16, 17]. The adrenal glands are massively enlarged with multiple, huge nodules that are typical, yellow-to-brown cortisol-producing adenomas (fig. 2). Most cases of MMAD are sporadic, although few familial cases have been described; in those, the disease appears in children. In some patients with MMAD, cortisol levels appear to increase with food ingestion (food-dependent CS) and in response to posture and other activities [16]. In these patients, aberrant expression of the neuroendocrine G-protein-coupled receptors (GPCRs) has been demonstrated in their adrenocortical tissue. Food-dependent CS has not been described in younger patients or in children, although bilateral macronodular

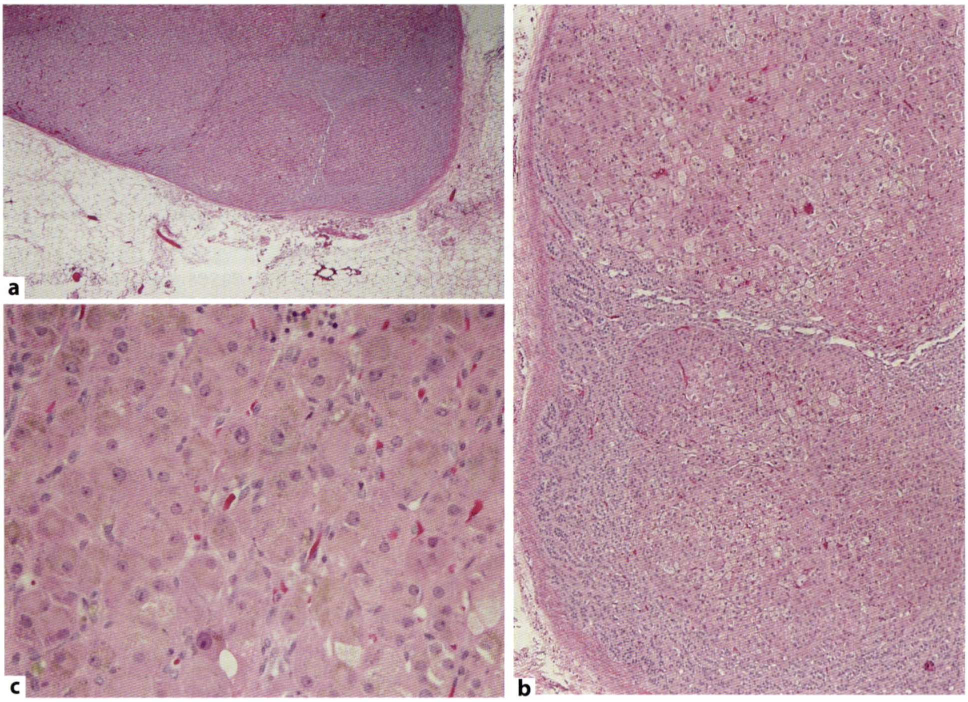

Fig. 4. **a–c** Lower (×5) and higher (×10 and ×40) magnification, respectively, of hematoxylin and eosin stainings of the patient whose adrenal is shown in figure 2d. The tissue demonstrates characteristic features of PPNAD such as multiple small nodules (**a**) surrounded by mostly atrophic or normal cortex (**b**) and cells that contain pigment (**c**) that is in most cases lipofuscin. Most PPNADs are due to *PRKAR1A* mutations but not all; isolated PPNAD (not associated with Carney complex) is frequently not associated with *PRKAR1A* mutations and this patient's *PRKAR1A* coding sequence was normal.

adrenal hyperplasia can also be seen in McCune-Albright syndrome [17]. In this syndrome there is a somatic mutation of the *GNAS* gene leading to constitutive activation of the Gsα protein and continuous, non-ACTH-dependent stimulation of the adrenal cortex. CS in MAS is rare and usually presents in the infantile period (before 6 months of age); interestingly, a few children have had spontaneous resolution of their CS [13, 14].

Because both PPNAD and MMAD and other BAHs can present with bilateral adrenal masses (fig. 2), a useful biochemical test is the 6-day-long Liddle test, modified to identify stimulation of UFC secretion, rather than suppression [18]. In this test, patients with micronodular forms of BAH respond with a gradual increase of UFC and 17-OHS secretion in response to the administration of dexamethasone (by 2 days of 0.5 mg dexamethasone every 6 h and 2 days of 2 mg dexamethasone every 6 h) (fig. 3). Although the cause of this 'paradoxical' rise in glucocorticoid synthesis by the adrenal cortex in response to dexamethasone is not known, it is a glucocorticoid receptor-mediated phenomenon [19] that may be due to the abnormal expression of various substances by the PPNAD cortex (such as synaptophysin and others, fig. 4) [20].

Adrenal Cancer and CS

Malignant neoplasias of the adrenal cortex account for 0.05–0.2% of all cancers, with an approximate prevalence of two new cases per million of population per year; adrenal cancer occurs at all ages, from early infancy to the 8th decade of life [21, 22]. A bimodal age of distribution has been reported, with the first peak occurring before the age of 5 years, and the second in the 4th to 5th decade. In all published series, females predominate, accounting for 65–90% of the reported cases. Several studies have shown a left-sided prevalence in adrenal cancer; however, others have reported a right-sided preponderance. In approximately 2–10% of the patients, adrenal cancer is found bilaterally. Overall, there appears to be a higher prevalence of adrenocortical carcinoma among patients with incidentally discovered adrenal masses than in the general population, although numerical estimates vary widely in the literature. Among the radiologically detectable masses, independent of size, one in 1,500 lesions may be an adrenal carcinoma; using the 5-cm cutoff as the most commonly accepted criterion for clinical investigation of an ADT, carcinoma may be found in as many as 7% of the patients with adrenal tumors over 5–7 cm in size [23].

In some areas of the world, higher incidence of adrenal cancer, especially in children, has been documented. This is particularly true for Southern Brazil, where environmental mutagens and a frequent *TP53* mutation have been postulated as the relevant pathogenic event [24]. In these areas, evaluation of incidentally discovered adrenal masses may be necessary for lesions smaller than 5 cm. Although the incidence of adrenal incidentalomas appears to be higher in some familial neoplasia syndromes like multiple endocrine neoplasia type 1 and familial adenomatous polyposis, it is unclear whether this finding is accompanied by a higher predisposition to adrenal cancer.

CS is most common among pediatric patients with adrenal carcinoma present with a hormonal syndrome which makes their detection easier and leads to their early surgical resection and medical treatment. CS is less frequent among adults with the disease.

Clinical and Molecular Genetics of ADTs Associated with CS

As we already mentioned, aberrant cAMP signaling has been linked to genetic forms of cortisol excess that lead to CS [25], mostly BAHs. Macronodular adrenocortical hyperplasia may be due to *GNAS* mutations associated with either McCune-Albright syndrome or sporadic ADTs. Micronodular BAH, and its better-known variant, PPNAD, may be caused by germline-inactivating mutations of the *PRKAR1A* gene [26]. Most patients with PPNAD also have Carney complex, as mentioned above [27].

Over the last several years, it has become apparent that there are several forms of micronodular BAH that are not caused by germline-inactivating mutations of the

PRKAR1A gene (table 1). We described one such case associated with an atypical, episodic, form of CS in a young child [14]. Her adrenal histology showed moderate diffuse cortical hyperplasia, multiple capsular deficits, and massive circumscribed and infiltrating extra-adrenal cortical excrescences that in many cases formed micronodules that were nonpigmented. Synaptophysin, a marker for PPNAD, also stained the nodules, in addition to the surrounding cortex [14, 20].

Recently, we reported that inactivating mutations of the *PDE11A* and *PDE8B* genes could be found in a subgroup of patients with PPNAD and other forms of BAH [28–30]. PDE11A is a dual-specificity phosphodiesterase catalyzing the hydrolysis of both cAMP and cGMP; it is expressed in several endocrine tissues, including the adrenal cortex [28, 29]. The *PDE11A* gene was mapped to the 2q31-35 chromosomal region and tumors from patients with PDE11A-inactivating mutations demonstrated 2q allelic losses (51). The *PDE11A* locus, like that of other PDEs, has a complex genomic organization; of the four possible splice variants, only A4 appears to be expressed in the adrenal cortex, whereas A1 is ubiquitous, and A2 and A3 have a more limited expression pattern. More recent data show that PDE11A is widely expressed in adrenocortical tissue and its expression appears to be modified in a variety of tumors beyond PPNAD and other forms of BAH.

PDE11A mutations and polymorphisms were found as low-penetrance predisposing factors to ADTs [29]. The *PDE11A* data support the notion that this gene is not necessarily causative of BAH but that it is associated with a low-penetrance predisposition to the development of BAH and possibly other ADTs leading to CS and, perhaps, other conditions.

More recently, a single *PDE8B* mutation was identified in a young child with BAH and CS [30]; PDE8B is another cAMP-specific PDE with wide expression in endocrine tissues, including the adrenal cortex. Its involvement in ADT formation beyond this rare case of isolated micronodular adrenocortical disease remains to be seen.

BAHs and cAMP Signaling

The cause of all forms of BAH studied to date appears to be linked to increased cAMP signaling. However, the histopathological changes in the adrenal glands of patients with the various mutations or functional abnormalities of this pathway differ significantly (fig. 2). *PRKAR1A* mutations are associated with the pigmented micronodular variant of BAH that is known as PPNAD, whereas *PDE11A* (and possibly PDE8B) mutations appear to be predisposing to a variety of lesions from isolated PPNAD to nonpigmented micronodular hyperplasia; *GNAS* mutations are associated with the macronodular and clearly nonpigmented forms of BAH. It is also interesting that sporadic ADTs (without any family history) can be associated with somatic mutations in all three of these genes.

It is noteworthy that in all forms of BAH associated with increased cAMP signaling there are patients with mutations in one of the causative genes that do not present with overt CS. The frequency with which carriers of mutations in one of these genes present with 'classic' CS appears to be higher in *PRKAR1A* mutation carriers than in *PDE11A, PDE8B,* or *GNAS*-associated disease with significant, however, interindividual variability and without a clear genotype-phenotype correlation. Interestingly, the age at which CS presents in these disorders is exactly the reverse, with Mc-Cune-Albright syndrome patients (*GNAS* mutation carriers) presenting almost always in infancy, whereas at least some of the patients with *PDE11A, PDE8B* mutations appear to present mostly in early childhood years and *PRKAR1A* mutation carriers in late adolescence and young adulthood. Thus, a number of factors are likely to affect the expression of these mutations, developmental, hormonal, and perhaps, gender-related ones. The presence of allelic losses of the corresponding normal allele in adrenal tissues seems to also be a determining factor in the development of disease associated with *PRKAR1A* and *PDE11A, PDE8B* mutations, since all these genes were identified using LOH studies [26, 28, 30].

PDE11A and *PDE8B* are the first PDEs to be linked to an inherited condition associated with tumor formation but may not be the only enzyme of this large family of proteins that predisposes to tumors. Our genome-wide dataset [28] suggests that other PDEs are likely to be involved in adrenal tumorigenesis in a similar manner: not by causing tumors per se, but by being a predisposing factor. Very little is known about PDE11A, PDE8B or other PDEs in adrenocortical tissue which, however, appears to exhibit significant PDE activity in vitro. Our preliminary data suggest that several PDEs are expressed in the cortex; *PDE11A* is expressed at levels that are higher than those of most other such enzymes with the exception of PDE8B [28, 30].

The high frequency of *PDE11A*-inactivating mutations in the population [29], the possibility that other members of this large family of proteins are involved in ADT formation, and the identification of clinically silent carriers [28–30] raise an interesting question: is it possible that *PDE11A* and *PDE8B* mutations (or mutations in a similar gene) underlie the high frequency of 'incidentalomas' [6] in the general population? At the moment, this question cannot be answered; larger and prospective studies need to be performed.

Surgical Treatment of ADTs Causing CS

Patients with benign ADTs are operated today mostly via a laparoscopic procedure (LP) that is preferred for both bilateral and unilateral lesions. LP has minimized morbidity and improvement is immediate after resection in these patients who are hypertensive preoperatively or have other complications of CS [31]. Replacement with glucocorticoids is necessary for up to 1 year after surgery for patients with unilateral adrenalectomies, whereas for patients after a bilateral procedure replacement with both gluco- and mineralocorticoids is necessary for life. For patients with cancer, the treatment of all primary tumors is also surgical, although open laparotomy for staging is preferable over LP. If complete resection of an adrenocortical carcinoma cannot be achieved, as much as possible of the tumor should be removed. Solitary recurrences or metastases should also be removed surgically, if possible. Long-term disease-free status has been produced by complete resection of adrenocortical carcinoma, whereas long-term remissions have followed surgical resection of hepatic, pulmonary, or cerebral metastases. Therapy with o,p'-DDD (mitotane) is initiated either as an adjuvant to surgical treatment or for patients with inoperable cancer [32, 33]. o,p'-DDD is an adrenocytolytic agent which is given at maximally tolerated oral doses (up 10 $g/m^2/day$). It ameliorates the endocrine syndrome in approximately two thirds, whereas tumor regression or arrest of growth has been observed in as many as one third of the patients. Occasionally, for the correction of hypercortisolism, steroid synthesis inhibitors (aminoglutethimide, metyrapone, trilostane, ketoconazole) or glucocorticoid antagonists (RU 486) are required. Patients taking mitotane (o,p'-DDD) may develop hypoaldosteronism or hypocortisolism, and fludrocortisone or hydrocortisone should be added as needed. Radiation therapy is occasionally helpful for palliation of metastases.

Concluding Remarks

CS caused by ADTs is most commonly caused by a solitary adenoma; BAHs are a more frequent cause of CS than previously thought. Defects of the cAMP signaling pathway are frequent in ADTs associated with CS. Cancer associated with CS is extremely rare. Surgical advances have made ADTs causing CS a disease that is cured in most cases with the exception of cancer.

Acknowledgments

This research was supported (in part) by the Intramural Research Program of the NIH/NICHD.

References

1 Orth DN: Cushing's syndrome. N Engl J Med 1995; 332:791–803.

2 Newell-Price JF, Trainer PF, Besser MF, Grossman AB: The diagnosis and differential diagnosis of Cushing's syndrome and pseudo-Cushing's states. Endocr Rev 1998;19:647–672.

3 Nieman LK: Diagnostic tests for Cushing's syndrome. Ann NY Acad Sci 2002;970:112–118.

4 Boscaro MF, Barzon LF, Fallo FF, Sonino N: Cushing's syndrome. Lancet 2001;357:783–791.

5 Normann T, Hanven J, Mjolnerod O: Cushing syndrome in an infant associated with neuroblastoma in two ectopic adrenal glands. J Pediatr Surg 1971; 6:169–175.

6 Saeger W, Reinhard K, Reinhard C: Hyperplastic and tumorous lesions of the adrenals in an unselected autopsy series. Endocr Pathol 1998;9:235–239.

7 Magiakou MA, Mastorakos G, Chrousos GP: Final stature in patients with endogenous Cushing's syndrome. J Clin Endocrinol Metab 1994;79:1082–1085.
8 Batista DL, Riar J, Keil M, Stratakis CA: Diagnostic tests for children who are referred for the investigation of Cushing syndrome. Pediatrics 2007;120: e575–e586.
9 Lebrethon MC, Grossman AB, Afshar FF, Plowman PN, Besser GM, Savage MO: Linear growth and final height after treatment for Cushing's disease in childhood. J Clin Endocrinol Metab 2000;85:3262–3265.
10 Papanicolaou DA, Yanovski JA, Cutler GB Jr, Chrousos GP, Nieman LK: A single midnight serum cortisol measurement distinguishes Cushing's syndrome from pseudo-Cushing states. J Clin Endocrinol Metab 1998;83:1163–1167.
11 Nieman LK, Oldfield EH, Wesley R, Chrousos GP, Loriaux DL, Cutler GB Jr: A simplified morning ovine corticotropin-releasing hormone stimulation test for the differential diagnosis of adrenocorticotropin-dependent Cushing's syndrome. J Clin Endocrinol Metab 1993;77:1308–1312.
12 Dichek HL, Nieman LK, Oldfield EH, Pass HI, Malley JD, Cutler GB Jr: A comparison of the standard high dose dexamethasone suppression test and the overnight 8-mg dexamethasone suppression test for the differential diagnosis of adrenocorticotropin-dependent Cushing's syndrome. J Clin Endocrinol Metab 1994;78:418–422.
13 Stratakis CA, Boikos SA: Genetics of adrenal tumors associated with Cushing's syndrome: a new classification for bilateral adrenocortical hyperplasias. Nat Clin Pract Endocrinol Metab 2007;3: 748–757.
14 Gunther DF, Bourdeau I, Matyakhina L, Cassarino D, Kleiner DE, Griffin K, Courkoutsakis N, Abu-Asab M, Tsokos M, Keil M, Carney JA, Stratakis CA: Cyclical Cushing syndrome presenting in infancy: an early form of primary pigmented nodular adrenocortical disease, or a new entity? J Clin Endocrinol Metab 2004;89:3173–3182.
15 Sarlis NJ, Chrousos GP, Doppman JL, Carney JA, Stratakis CA: Primary pigmented nodular adrenocortical disease: reevaluation of a patient with carney complex 27 years after unilateral adrenalectomy. J Clin Endocrinol Metab 1997;82:1274–1278.
16 Bourdeau I, Lampron A, Costa MH, Tadjine M, Lacroix A: Adrenocorticotropic hormone-independent Cushing's syndrome. Curr Opin Endocrinol Diabetes Obes 2007;14:219–225.
17 Stratakis CA, Kirschner LS: Clinical and genetic analysis of primary bilateral adrenal diseases (micro- and macronodular disease) leading to Cushing syndrome. Horm Metab Res 1998;30:456–463.
18 Stratakis CA, Sarlis N, Kirschner LS, Carney JA, Doppman JL, Nieman LK, Chrousos GP, Papanicolaou DA: Paradoxical response to dexamethasone in the diagnosis of primary pigmented nodular adrenocortical disease. Ann Intern Med 1999;131: 585–591.
19 Bourdeau I, Lacroix A, Schürch W, Caron P, Antakly T, Stratakis CA: Primary pigmented nodular adrenocortical disease: paradoxical responses of cortisol secretion to dexamethasone occur in vitro and are associated with increased expression of the glucocorticoid receptor. J Clin Endocrinol Metab 2003;88:3931–3937.
20 Stratakis CA, Carney JA, Kirschner LS, Willenberg HS, Brauer S, Ehrhart-Bornstein M, Bornstein SR: Synaptophysin immunoreactivity in primary pigmented nodular adrenocortical disease: neuroendocrine properties of tumors associated with Carney complex. J Clin Endocrinol Metab 1999;84:1122–1128.
21 Allolio B, Hahner S, Weismann D, Fassnacht M: Management of adrenocortical carcinoma. Clin Endocrinol (Oxf) 2004;60:273–287.
22 Libè R, Fratticci A, Bertherat J: Adrenocortical cancer: pathophysiology and clinical management. Endocr Relat Cancer 2007;14:13–28.
23 Grumbach MM, Biller BM, Braunstein GD, Campbell KK, Carney JA, Godley PA, Harris EL, Lee JK, Oertel YC, Posner MC, Schlechte JA, Wieand HS: Management of the clinically inapparent adrenal mass ('incidentaloma'). Ann Intern Med 2003;138: 424–429.
24 Ribeiro RC, Sandrini F, Figueiredo B, Zambetti GP, Michalkiewicz E, Lafferty AR, DeLacerda L, Rabin M, Cadwell C, Sampaio G, Cat I, Stratakis CA, Sandrini R: An inherited p53 mutation that contributes in a tissue-specific manner to pediatric adrenal cortical carcinoma. Proc Natl Acad Sci USA 2001;98:9330–9335.
25 Stratakis CA: Genetics of adrenocortical tumors: gatekeepers, landscapers and conductors in symphony. Trends Endocrinol Metab 2003;14:404–410.
26 Kirschner LS, Carney JA, Pack SD, Taymans SE, Giatzakis C, Cho YS, Cho-Chung YS, Stratakis CA: Mutations of the gene encoding the protein kinase A type I-alpha regulatory subunit in patients with the Carney complex. Nat Genet 2000;26:89–92.
27 Stratakis CA, Kirschner LS, Carney JA: Clinical and molecular features of the Carney complex: diagnostic criteria and recommendations for patient evaluation. J Clin Endocrinol Metab 2001;86:4041–4046.

28 Horvath A, Boikos S, Giatzakis C, Robinson-White A, Groussin L, Griffin KJ, Stein E, Levine E, Delimpasi G, Hsiao HP, Keil M, Heyerdahl S, Matyakhina L, Libè R, Fratticci A, Kirschner LS, Cramer K, Gaillard RC, Bertagna X, Carney JA, Bertherat J, Bossis I, Stratakis CA: A genome-wide scan identifies mutations in the gene encoding phosphodiesterase 11A4 (PDE11A) in individuals with adrenocortical hyperplasia. Nat Genet 2006;38:794–800.

29 Horvath A, Giatzakis C, Robinson-White A, Boikos S, Levine E, Griffin K, Stein E, Kamvissi V, Soni P, Bossis I, de Herder W, Carney JA, Bertherat J, Gregersen PK, Remmers EF, Stratakis CA: Adrenal hyperplasia and adenomas are associated with inhibition of phosphodiesterase 11A in carriers of PDE11A sequence variants that are frequent in the population. Cancer Res 2006;66:11571–11575.

30 Horvath A, Mericq V, Stratakis CA: Mutation in PDE8B, a cAMP-specific phosphodiesterase in adrenal hyperplasia. N Engl J Med 2008;358:750–752.

31 Gallagher SF, Wahi M, Haines KL, Baksh K, Enriquez J, Lee TM, Murr MM, Fabri PJ: Trends in adrenalectomy rates, indications, and physician volume: a statewide analysis of 1816 adrenalectomies. Surgery 2007;142:1011–1021.

32 Schteingart DE: Adjuvant mitotane therapy of adrenal cancer – use and controversy. N Engl J Med 2007;356:2415–2418.

33 Terzolo M, Angeli A, Fassnacht M, Daffara F, Tauchmanova L, Conton PA, Rossetto R, Buci L, Sperone P, Grossrubatscher E, Reimondo G, Bollito E, Papotti M, Saeger W, Hahner S, Koschker AC, Arvat E, Ambrosi B, Loli P, Lombardi G, Mannelli M, Bruzzi P, Mantero F, Allolio B, Dogliotti L, Berruti A: Adjuvant mitotane treatment for adrenocortical carcinoma. N Engl J Med 2007;356:2372–2380.

Constantine A. Stratakis
Section on Endocrinology and Genetics
Program on Developmental Endocrinology and Genetics, National Institute of Child Health and Human Development, National Institutes of Health
Building 10, CRC, Room I-3330, 10 Center Dr., MSC 1103, Bethesda, MD 20892 (USA)
Tel. +1 301 496 46 86, Fax +1 301 402 0574, E-Mail stratakc@mail.nih.gov

Flück CE, Miller WL (eds): Disorders of the Human Adrenal Cortex.
Endocr Dev. Basel, Karger, 2008, vol 13, pp 133–144

The Role of Adrenal Steroidogenesis in Arterial Hypertension

Markus G. Mohaupt

Division of Hypertension, Department of Nephrology/Hypertension, University of Bern, Berne, Switzerland

Abstract

Adrenal aldosterone production, the major regulator of salt and water retention, is discussed with respect to hypertensive diseases. Physiological aldosterone production is tightly regulated, either stimulated or inhibited, in the adrenal zona glomerulosa by both circulating factors and/or by locally derived endothelial factors. Arterial hypertension caused by volume overload is the leading clinical symptom indicating increased mineralocorticoid hormones. Excessive aldosterone production is seen in adenomatous disease of the adrenals. The balance between stimulatory/proliferative and antagonistic signaling is disturbed by expression of altered receptor subtypes in the adenomas. Increased aldosterone production without a detectable adenoma is the most frequent form of primary aldosteronism. Both increased sensitivity to agonistic signals and activating polymorphisms within the aldosterone synthase gene *(CYP11B2)* have been associated with excessive aldosterone production. 17α-Hydroxylase deficiency and glucocorticoid-remediable aldosteronism can also cause excessive mineralocorticoid synthesis. In contrast, the severe form of pregnancy-induced hypertension, preeclampsia, is characterized by a compromised volume expansion in the presence of inappropriately low aldosterone levels. Initial evidence suggests that compromised CYP11B2 is causative, and that administration of NaCl lowered blood pressure in pregnant patients with low aldosterone availability due to a loss of function.

Adrenal aldosterone production is a major regulator affecting salt and water retention in the distal tubule and cortical collecting ducts of the kidneys. Severe cases of aldosterone overproduction (Conn's disease) present a clinical picture with severe hypokalemia, arterial hypertension, metabolic alkalosis and adrenal adenoma. Using the aldosterone-renin ratio to identify autonomic production of aldosterone, the clinical picture is less obvious with a lower prevalence of hypokalemia. Thus, this ratio is an insensitive predictor of the disease and the number of adrenal adenomas is also substantially reduced. In contrast to other disorders of adrenal steroid hormone synthesis with either reduced aldosterone production (such as 3β-hydroxysteroid dehydrogenase or 21-hydroxylase deficiencies) or increased aldosterone production (such as a 17α-hydroxylase deficiency), where there are also disorders of sexual differentiation, the principal symptom of primary aldosteronism is arterial hypertension, which is frequently not recognized.

Several converging pathways interact in a complex regulatory interplay of stimulatory and inhibitory factors acting on the zona glomerulosa. In contrast to arterial hypertension reflecting volume overload, in pregnancy compromised plasma volume is associated with hypertension and preeclampsia. Initial evidence suggests that compromised availability of aldosterone contributes to this disease.

Regulation of Aldosterone Biosynthesis

Aldosterone is produced within the zona glomerulosa by the enzyme aldosterone synthase (CYP11B2). The gene encoding the CYP11B2 enzyme has 93% identity in its exons and 90% identity in its introns with the gene for 11β-hydroxylase (CYP11B1), suggesting an ancestral gene duplication [1, 2]. Thus, while the coding regions differ by only 35 amino acids, the 5′-upstream region of the *CYP11B2* gene diverges considerably from that of the *CYP11B1* gene. Unlike the rat, the human has only two such genes mapping to the long arm of chromosome 8 approximately 40 kb apart [3, 4]. Both genes consist of 9 exons and 8 introns.

Aldosterone Synthesis

Aldosterone synthesis differs from the production of sex steroids, cortisol and corticosterone by its localization in the zona glomerulosa and by the involvement of a specific enzyme, CYP11B2, which is located on the matrix side of the inner mitochondrial membrane. The development of selective inhibitors added to the understanding of structure-function relationships, which still require structural modeling based on various CYP structures from different microorganisms, since no X-ray structure analysis is available for the mitochondrial CYP11B2 [5, 6]. Though highly homologous, the CYP11B1 displays 11β-hydroxylase capabilities, but only marginal 18-hydroxylase activity. CYP11B2 possesses 11β- and 18-hydroxylase activity. The rate-limiting step of aldosterone production specific for the CYP11B2 enzyme is the 18-methyl oxidase activity [7].

Cellular Mechanisms Regulating Aldosterone Production

Major stimuli to enhance aldosterone production under normal conditions are angiotensin II, potassium, low sodium, and stress [8, 9]. The primary stimulator of aldosterone synthesis is the renin-angiotensin system, which is activated by changes in blood sodium concentrations [10] and blood pressure; these changes are sensed by the macula densa and juxtaglomerular cells. Multiple signals regulate aldosterone synthesis, including endothelin, adrenomedullin, nitric oxide, catecholamines, ACTH, vasopressin, serotonin, interleukin-1, neuromedin N, hypocretin, orexin, and potassium ions. Major endogenous inhibitors appear to be dopamine, atrial natriuretic hormone, and nitric oxide [11, 12]. Some of these mechanisms are overly expressed and thus recognizable in adenomatous disease (Conn syndrome), where they interfere with CYP11B2 expression and aldosterone production (discussed below) [13]. In addition to these regulators, activin and bone morphogenetic protein-6 (BMP-6) play important roles in enhancing transcription of the genes for the steroidogenic acute regulatory protein (StAR) and CYP11B2. The latter augmented angiotensin II-dependent aldosterone secretion, but not ACTH- and potassium-related activation [14, 15]. In contrast, angiotensin II and aldosterone suppress BMP-6 in an autoregulatory feedback mode, suggesting that BMP-6 may be a key autocrine modulator for aldosterone production [16].

Adrenal vascularization is high and thus the interaction between endothelial and adrenal cells is intense [17]. Conditioning cultured human adrenal cells with supernatant from endothelial cells enhanced aldosterone production via increased expression of the mRNAs for CYP11B2 and StAR [18, 19]. Beyond basal rates, StAR promoter activity is induced by cAMP via a cAMP-response element modulator [20, 21].

Arterial Hypertension in the Presence of an Altered Production of Mineralocorticoids

Several diseases can lead to increased adrenal aldosterone production. These include autonomous aldosterone production in adrenal tissue with or without overt adenomas, 17α-hydroxylase deficiency with hypertension induced by deoxycorticosterone and amplified expression of StAR protein, or enhanced availability of atypical mineralocorticoids, as is seen in glucocorticoid-remediable aldosteronism.

Prevalence of Primary Aldosteronism

Earlier estimates of the incidence of primary aldosteronism suggested that its prevalence was less than 2%. Hypokalemia was the leading clinical feature used to diagnose inappropriate aldosterone secretion, and normokalemia was used to exclude this syndrome. Current observations based on hormonal data propose that hypokalemia is present in less than 50% of patients, either with or without adrenal adenomas, although adenomas present slightly more frequently with hypokalemia [22, 23]. Due to the different approaches formerly used to diagnose primary aldosteronism, whether the incidence of primary aldosteronism is increasing remains a matter of debate, irrespective of its pathogenesis. Current studies in a general population of hypertensive patients, thus avoiding referral bias, indicate an incidence of 13–14.4% of affected individuals [24]. The prevalence is rising with the severity of arterial hypertension [25].

Adenoma in Primary Aldosteronism

Aldosterone-producing adenomas meet the definition of an autonomous production with little or no regulatory responses. It has been proposed that the diagnosis of an adenoma requires an increase in aldosterone, and (perhaps more reliably) of its precursor 18-OH-corticosterone following a change from the supine to the upright posture, but this lacks diagnostic sensitivity [26]. The role of 18-OH-corticosterone is questioned by the recent finding that the aldosterone/cortisol ratio is more diagnostic than the 18-OH-corticosterone/cortisol ratio in adrenal vein sampling [27]. Surprisingly, advanced imaging techniques, including NMR and CT scans, were unable to predict lateralization of aldosterone overproduction reliably when compared to adrenal vein sampling [28]. Interestingly, in a broad-based Italian cohort involving 1,125 hypertensive individuals following hormonal diagnosis of primary aldosteronism, lateralization studies and adrenalectomy were performed and the clinical outcome was tracked. Adenomas were confirmed in 43% of the patients, constituting 4.8% of all hypertensive patients [23].

Angiotensin II is the major regulator of aldosterone production under normal conditions. There are two angiotensin II receptor subtypes, both expressed in adrenal tissue. Proliferation and aldosterone production are principally increased via angiotensin II subtype 1 receptors, yet both subtypes seem to contribute [29]. The normal angiotensin II response is enhanced in functional adenomas, which have a higher angiotensin II subtype 1 receptor expression compared to nonfunctional adenomas [30]. In addition to the higher expression of agonistic angiotensin II subtype 1 receptors, increased intracellular signals such as a higher cyclic ADP-ribosyl cyclase activity enhance aldosterone production in adenomas [31]. Angiotensin II-dependent activation of aldosterone production is also potentiated by dopamine type 4 receptors and reduced by type 2 receptor activation, contrasting with the otherwise aldosterone-antagonistic properties of dopamine signaling [32]. A mechanism adding to aldosterone overproduction in adenomas, like the downregulation of attenuating receptors in other endocrine tumors, is a reduced expression of dopamine type 2 receptors. This was associated with an increased protein kinase C-mu phosphorylation and membrane translocation resulting in an overall increase in CYP11B2 mRNA expression and aldosterone secretion [33]. Other vasoconstrictor mechanisms may also be altered in adenomas, such as the expression of serotonin 4 receptors [34] and functionally supporting the activation of vasopressin V_{1a} receptors [35].

Nonadenomatous Excess of Mineralocorticoid Active Hormones

Idiopathic aldosterone production seems to follow a pathophysiological cascade that includes increased angiotensin II sensitivity and thus more resembles secondary hyperaldosteronism. Increased sensitivity of the adrenals might be mediated by enhanced cholesterol delivery to the steroidogenic enzymes, as has been shown in the presence of ACTH in zona glomerulosa cells [36]. Obesity and hypertension are clinically closely related. Steroidogenic factor 1 (SF-1) sites are key transcription factors of *CYP11B2*. Wnt signaling and the activation of SF-1 sites have been demonstrated to have the potency to converge [37]. Adipocytes secrete Wnt-signaling molecules, which act via cell surface receptors. Accordingly, Wnt- and SF-1-mediated transcription has been observed in response to adipocyte-conditioned medium, which resulted in an increased aldosterone synthesis in human adrenal cells [38]. Similarly, adipocyte-conditioned medium induced ERK1/2 MAPK-mediated upregulation of StAR, sensitizing adrenal cells to angiotensin II-dependent aldosterone production [39].

Two linked mutations in the gene coding for CYP11B2 have been associated with arterial hypertension and an elevated aldosterone to renin ratio: first, the −344C/T polymorphism in the SF-1 binding site of the promotor region; second, the *CYP11B2* to *CYP11B1* conversion in intron 2 [40, 41]. An increased production of deoxycorticosterone, rarely related to an atypical steroid hormone-producing adrenal carcinoma, is found in 17α-hydroxylase deficiency, where 17-deoxysteroids in the aldosterone-biosynthetic pathway are overproduced [42, 43]. Glucocorticoid-remediable aldosteronism is the consequence of an unequal crossover between *CYP11B1* and *CYP11B2*. The resulting hybrid gene has the *CYP11B1* sequence including the ACTH-sensitive promotor region at the 5′ end, and the *CYP11B2* region at the 3′ end. Consequently, high 18-hydroxycortisol and aldosterone levels in the presence of rather low cortisol levels are frequently observed. Recently, even in the presence of elevated 18-hydroxycortisol concentrations glucorticoid-remediable aldosteronism was excluded as contributing to arterial hypertension [44].

Hypertension in Pregnancy and Aldosterone Deficiency

Volume Status and Blood Pressure in Pregnancy

The high aldosterone concentrations and increased intravascular volume found in normal pregnancies represent a physiological adaptation that is teleologicaly meaningful for placental perfusion and fetal substrate delivery [45–47]. Depletion of intravascular volume characterizes severe pregnancy-associated disease states such as intrauterine growth retardation, chronic hypertension or preeclampsia [45]. Two hypotheses have been suggested to explain volume depletion in pregnancy. The first ascribes inappropriate sensing of vascular 'overfilling' resulting in an increased transendothelial loss of fluid to the extravascular compartment. By contrast, the second hypothesis focuses on vascular 'underfilling' due to inappropriately low aldosterone levels. The second hypothesis is based on the assumption that a compensatory increase in the circulating fluid volume is required in normal pregnancy to support fetal substrate delivery. According to the second concept, maternal blood pressure increases in preeclampsia/hypertension due to counterregulatory mechanisms initiated by a reduced placental blood supply [48]. Observations that compromised volume status before pregnancy predicts a complicated pregnancy outcome are in support of the 'underfilling' hypothesis [46]. Earlier experimental studies suggested that those who develop preeclampsia had a reduced ability to retain sodium [47]. In women destined to develop preeclampsia, plasma volume reduction is often accompanied by an increased diastolic blood pressure [45]. The relevance of intravascular volume expansion is also supported by clinical observations that reduction of fluids by diuretics or salt restriction does not prevent or improve the course of the disease. On the contrary, acute volume expansion in overt preeclampsia can transiently decrease blood pressure. The relevance of volume expansion for fetal development is supported by investigations performed in pregnant sheep [49].

The role of volume expansion from sodium chloride intake during pregnancy was suggested by Robinson [50], who found a reduced incidence of preeclampsia in pregnant women on a high-salt diet. This study found that the chances of developing preeclampsia were reduced (relative risk 0.45, 95% confidence interval 0.26–0.72) by consuming 19–20 g dietary sodium chloride per day. The study design in this and two other studies on a high-salt diet during pregnancy (performed 40 years ago) was potentially biased [51]. In contrast, a Cochrane meta-analysis of two studies on a low- versus normal-salt diet in pregnancy to prevent preeclampsia including 603 patients indicated an unchanged relative risk to develop preeclampsia of 1.11 [52]. Low aldosterone production was not considered in these early studies due to the absence of appropriate techniques for genetic and hormonal analysis.

Aldosterone Synthesis in Pregnancies without and with Preeclampsia

Based on the 'underfilling' theory, compromised aldosterone synthesis would reduce the pregnancy-associated expansion of intravascular volume leading to inadequate placental perfusion and placental ischemia, which might induce systemic maternal hypertension leading to the clinical picture of preeclampsia. The hypothesis that reduced aldosterone synthesis accounts for preeclampsia in some patients is supported by diminished aldosterone production at 26–29 weeks of gestation in preeclampsia [53]. Understanding the genetics of the enzymes responsible for aldosterone synthesis gave a clue to the relevance of the blunted increase in aldosterone synthesis for the pathogenesis of preeclampsia. The apparently rate-limiting activity of the 18-methyl oxidase activity of aldosterone synthase was diminished in preeclampsia [54]. In this study, steroid hormones with mineralocorticoid activity, such as deoxycorticosterone or corticosterone, did not substitute for the missing aldosterone in preeclamptic pregnancies. In addition, other observations suggested that deoxycorticosterone was unresponsive to variations in sodium loading in normal pregnancies, indicating that this steroid is unlikely to substitute for the missing aldosterone [55].

Genetics of CYP11B2 in Preeclampsia

Several point mutations cause congenital corticosterone methyloxidase deficiency (CMO) leading to hypoaldosteronism with variable clinical symptoms. Two syndromes have been described. CMO type I (CMO I) is characterized by high concentrations of corticosterone and undetectable concentrations of 18-OH-corticosterone and aldosterone, whereas CMO type II (CMO II) is characterized by high concentrations of both corticosterone and 18-OH-corticosterone and undetectable concentrations of aldosterone [56]. CMO II has been described with the amino acid changes V386A, R181W, R173K, T318M and Q198D alone or in combination [6, 57–60]. The mutation R173K aggravates reduced CYP11B2 activity in the presence of further mutations in transfection assays [61]. CMO II deficiency is frequent among Jews originating from the Iran (1 in 4,000 births) presenting with the missense mutations R181W and V386A. No information on the course of pregnancy in these patients has been reported [59, 62]. The severe form (CMO I) has been found with gene deletions, amino acid insertions or amino acid changes at critical loci, such as a 5-bp deletion of exon 1, an L461P or E255X transition, and an insertion of two amino acids at exon 3 [63–66].

A stimulus for CYP11B2 transcription and increased aldosterone production in pregnancy appears to be progesterone [67]. Studies in pregnant rats show increased gestational expression of CYP11B2, which is enhanced by a low-salt diet [7, 68, 69]. Its augmented production suggests that increased amounts of aldosterone are required in normal pregnancy to accommodate gestational volume homeostasis. The human aldosterone/renin ratio increases manyfold and the aldosterone production is much higher in pregnant than in nonpregnant women [70]. The underlying mechanisms are a matter of debate.

Our group found no CYP11B2 sequence variants in 50% of women with regular pregnancies, but detected such variants in 76% of preeclamptic women [54]. The polymorphism V386A was found only in preeclamptic women, affecting 17% of these women. In a French control population, this mutation was seen in 4% of normotensive patients and in 18% of patients with essential hypertension, but this study provided no information about pregnancy [6]. The V386A mutation has also been described in patients with mild CMO II deficiency [6, 57, 59, 71]. In vitro transfection studies suggest that A386V has a slightly reduced capacity to produce aldosterone [6, 59]. V386A is a sequence normally present in CYP11B1, hence it may be either an ancestral gene conversion or an independent mutation [59]. In combination with other mutations, which do not alter CYP11B2 activity significantly themselves, the polymorphism V386A further reduced the CYP11B2 activity [6]. Mild CMO II deficiencies may be rather frequent and exacerbated at times of sodium deprivation [59]. It is conceivable that pregnancy due to its increased demand for aldosterone reveals mild CMO deficiencies.

These observations are of potential clinical relevance. A pregnant woman homozygous for the V386A polymorphism with borderline hypertension during the first trimester had consistently low apparent 18-methyl oxidase activity and low urinary tetrahydroaldosterone excretion. She was supplemented with sodium chloride throughout pregnancy, and mean systolic and diastolic blood pressure dropped by 16 and 12 mm Hg, respectively (fig. 1). Another woman with a similar genetic background, who experienced severe preeclampsia in a previous pregnancy and was advised to follow a low-salt diet in another hospital, also had an uneventful course of her 2nd pregnancy.

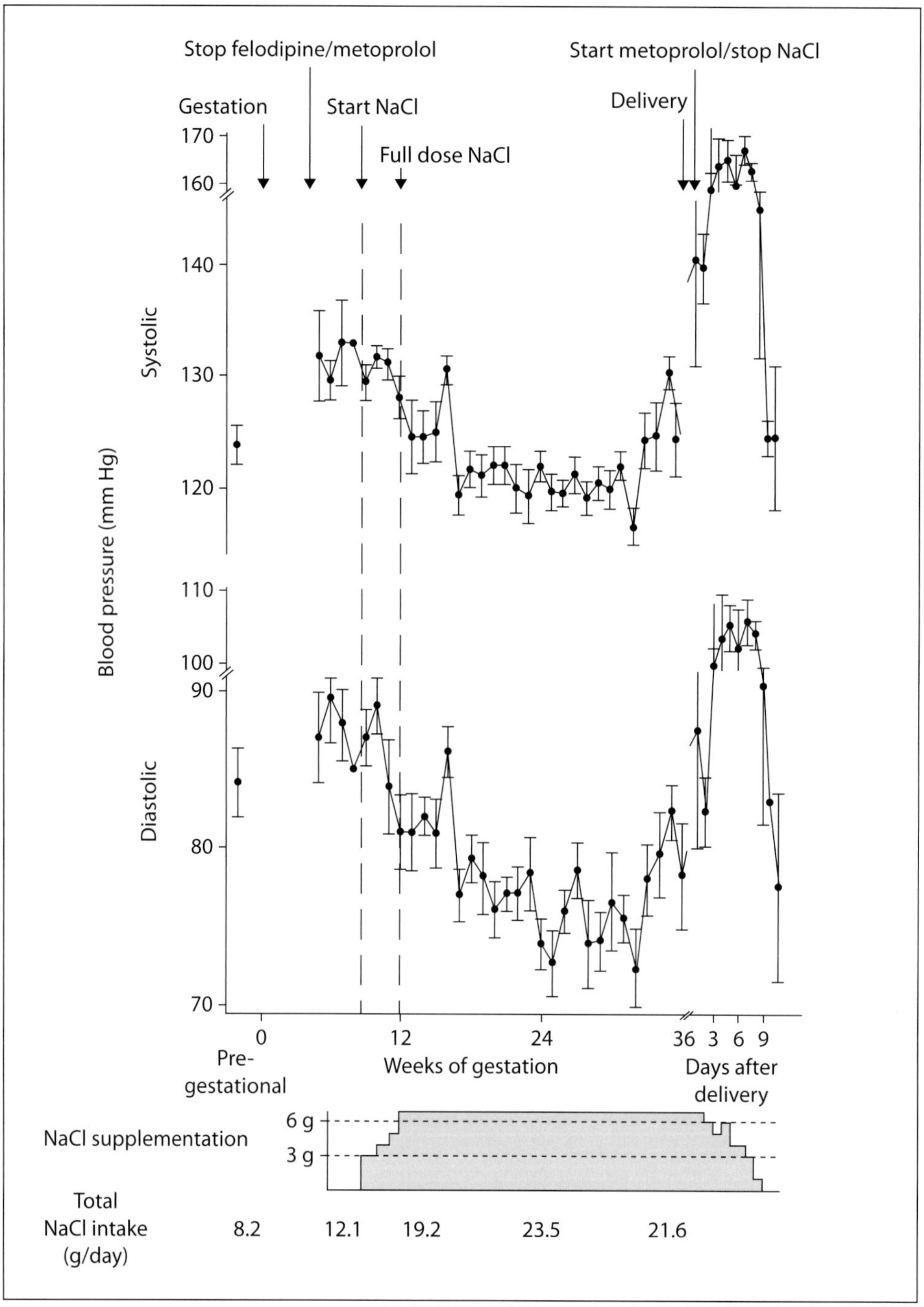

Fig. 1. Systolic and diastolic blood pressure is presented. Pregestational blood pressure is the mean ± SEM of monthly office cuff measurements during 6 preceding months. Gestational blood pressure is presented as weekly mean ± SEM of triplicate daily measurements verified by office measurements every other week. After delivery, daily means ± SEM of three to five measurements are given. Sodium chloride supplementation is indicated at the bottom. Total (supplemental and dietary) sodium chloride uptake as calculated from 24-hour urinary Na^+ excretion in g/day [reprinted from 74].

These observations in women with mild forms of CMO II deficiency support the assumption of supplementing salt to account for reduced aldosterone availability in pregnancy, a state of physiological volume expansion. An appropriate plasma volume will then provide for an adequate uteroplacental perfusion. Antiangiogenetic factors, such as sFlt-1, reduce maternal vasodilatation and are released by the placenta in response to hypoxia [72, 73]. Consequently, improving placental perfusion is expected to reduce hypoxia-induced, placental antiangiogenetic signals and thus secondarily maternal blood pressure.

Acknowledgements

I acknowledge the Swiss National Foundation (No. 3200B0-113902/1) for their support.

References

1 Mornet E, Dupont J, Vitek A, White PC: Characterization of two genes encoding human steroid 11β-hydroxylase (P-$450_{11\beta}$). J Biol Chem 1989;264: 20961–20967.

2 Kawamoto T, Mitsuuchi Y, Toda K, Yokoyama Y, Miyahara K, Miura S, Ohnishi T, Ichikawa Y, Nakao K, Imura H, Ulick S, Shizuta Y: Role of steroid 11β-hydroxylase and steroid 18-hydroxylase in the biosynthesis of glucocorticoids and mineralocorticoids in humans. Proc Natl Acad Sci USA 1992;89: 1458–1462.

3 Taymans SE, Pack S, Pak E, Torpy DJ, Zhuang Z, Stratakis CA: Human CYP11B2 (aldosterone synthase) maps to chromosome 8q24.3. J Clin Endocrinol Metab 1998;83:1033–1036.

4 Zhang G, Miller WL: The human genome contains only two CYP11B (P450c11) genes. J Clin Endocrinol Metab 1996;81:3254–3256.

5 Bechtel S, Belkina N, Bernhardt R: The effect of amino-acid substitutions I112P, D147E and K152N in CYP11B2 on the catalytic activities of the enzyme. Eur J Biochem 2002;269:1118–1127.

6 Portrat-Doyen S, Tourniaire J, Richard O, Mulatero P, Aupetit-Faisant B, Curnow KM, Pascoe L, Morel Y: Isolated aldosterone synthase deficiency caused by simultaneous E198D and V386A mutations in the CYP11B2 gene. J Clin Endocrinol Metab 1998;83: 4156–4161.

7 Malee MP, Mellon SH: Zone-specific regulation of two messenger RNAs for P450c11 in the adrenals of pregnant and nonpregnant rats. Proc Natl Acad Sci USA 1991;88:4731–4735.

8 Kakiki M, Morohashi K, Nomura M, Omura T, Horie T: Regulation of aldosterone synthase cytochrome P450 (CYP11B2) and 118-hydroxylase cytochrome P450 (CYP11B1) expression in rat adrenal zona glomerulosa cells by low sodium diet and angiotensin II receptor antagonists. Biol Pharm Bull 1997; 20:962–968.

9 LeHoux JG, Lefebvre A: Transcriptional activity of the hamster CYP11B2 promoter in NCI-H295 cells stimulated by angiotensin II, potassium, forskolin and bisindolylmaleimide. J Mol Endocrinol 1998;20: 183–191.

10 Binnion PF, Davis JO, Brown TC, Olichney MJ: Mechanisms regulating aldosterone secretion during sodium depletion. Am J Physiol 1965;208:655–661.

11 McKenna TJ, Island DP, Nicholson WE, Liddle GW: Dopamine inhibits angiotensin-stimulated aldosterone biosynthesis in bovine adrenal cells. J Clin Invest 1979;64:287–291.

12 Hanke CJ, Drewett JG, Myers CR, Campbell WB: Nitric oxide inhibits aldosterone synthesis by a guanylyl cyclase-independent effect. Endocrinology 1998;139:4053–4060.

13 Ehrhart-Bornstein M, Hinson JP, Bornstein SR, Scherbaum WA, Vinson GP: Intraadrenal interactions in the regulation of adrenocortical steroidogenesis. Endocr Rev 1998;19:101–143.

14 Suzuki J, Otsuka F, Inagaki K, Takeda M, Ogura T, Makino H: Novel action of activin and bone morphogenetic protein in regulating aldosterone production by human adrenocortical cells. Endocrinology 2004;145:639–649.

15 Inagaki K, Otsuka F, Suzuki J, Kano Y, Takeda M, Miyoshi T, Otani H, Mimura Y, Ogura T, Makino H: Involvement of bone morphogenetic protein-6 in differential regulation of aldosterone production by angiotensin II and potassium in human adrenocortical cells. Endocrinology 2006;147:2681–2689.
16 Inagaki K, Otsuka F, Suzuki J, Otani H, Takeda M, Kano Y, Miyoshi T, Yamashita M, Ogura T, Makino H: Regulatory expression of bone morphogenetic protein-6 system in aldosterone production by human adrenocortical cells. Regul Pept 2007;138:133–140.
17 Vinson GP, Pudney JA, Whitehouse BJ: The mammalian adrenal circulation and the relationship between adrenal blood flow and steroidogenesis. J Endocrinol 1985;105:285–294.
18 Ansurudeen I, Kopprasch S, Ehrhart-Bornstein M, Willenberg HS, Krug AW, Funk RHW, Bornstein SR: Vascular-adrenal niche – endothelial cell-mediated sensitization of human adrenocortical cells to angiotensin II. Horm Metab Res 2006;38:476–480.
19 Ansurudeen I, Kopprasch S, Ehrhart-Bornstein M, Bornstein SR, Willenberg HS: Endothelial cell-mediated regulation of aldosterone release from human adrenocortical cells. Mol Cell Endocrinol 2007;265–266:150–156.
20 Clark BJ, Combs R: Angiotensin II and cyclic adenosine 3′,5′-monophosphate induce human steroidogenic acute regulatory protein transcription through a common steroidogenic factor-1 element. Endocrinology 1999;140:4390–4398.
21 Sugawara T, Sakuragi N, Minakami H: CREM confers cAMP responsiveness in human steroidogenic acute regulatory protein expression in NCI-H295R cells rather than SF-1/Ad4BP. J Endocrinol 2006;191: 327–337.
22 Mosso L, Carvajal C, González A, Barraza A, Avila F, Montero J, Huete A, Gederlini A, Fardella CE: Primary aldosteronism and hypertensive disease. Hypertension 2003;42:161–165.
23 Rossi GP, Bernini G, Caliumi C, Desideri G, Fabris B, Ferri C, Ganzaroli C, Giacchetti G, Letizia C, Maccario M, Mallamaci F, Mannelli M, Mattarello M-J, Moretti A, Palumbo G, Parenti G, Porteri E, Semplicini A, Rizzoni D, Rossi E, Boscaro M, Pessina AC, Mantero F, PAPY study investigators: A prospective study of the prevalence of primary aldosteronism in 1,125 hypertensive patients. J Am Coll Cardiol 2006; 48:2293–2300.
24 Lim PO, Rodgers P, Cardale K, Watson AD, MacDonald TM: Potentially high prevalence of primary aldosteronism in a primary-care population. Lancet 1999;353:40.
25 Calhoun DA: Aldosteronism and hypertension. Clin J Am Soc Nephrol 2006;1:1039–1045.
26 Biglieri EG, Schambelan M: The significance of elevated levels of plasma 18-hydroxycorticosterone in patients with primary aldosteronism. J Clin Endocrinol Metab 1979;49:87–91.
27 Auchus RJ, Chandler DW, Singeetham S, Chokshi N, Nwariaku FE, Dolmatch BL, Holt SA, Wians FH Jr, Josephs SC, Trimmer CK, Lopera J, Vongpatanasin W, Nesbitt SD, Leonard D, Victor RG: Measurement of 18-hydroxycorticosterone during adrenal vein sampling for primary aldosteronism. J Clin Endocrinol Metab 2007;92:2648–2651.
28 Nwariaku FE, Miller BS, Auchus R, Holt S, Watumull L, Dolmatch B, Nesbitt S, Vongpatanasin W, Victor R, Wians F, Livingston E, Snyder WH: Primary hyperaldosteronism: effect of adrenal vein sampling on surgical outcome. Arch Surg 2006;141:497–503.
29 Tanabe A, Naruse M, Arai K, Naruse K, Yoshimoto T, Seki T, Imaki T, Kobayashi M, Miyazaki H, Demura H: Angiotensin II stimulates both aldosterone secretion and DNA synthesis via type 1 but not type 2 receptors in bovine adrenocortical cells. J Endocrinol Invest 1998;21:668–672.
30 Schubert B, Fassnacht M, Beuschlein F, Zenkert S, Allolio B, Reincke M: Angiotensin II type 1 receptor and ACTH receptor expression in human adrenocortical neoplasms. Clin Endocrinol 2001;54:627–632.
31 Takeda Y, Usukura M, Yoneda T, Oda N, Ito Y, Mabuchi H: The expression of messenger RNA for ADP-ribosyl cyclase in aldosterone-producing adenomas. Clin Endocrinol 2005;62:504–508.
32 Wu K-D, Chen Y-M, Chu T-S, Chueh S-C, Wu M-H, Hsieh B-S: Expression and localization of human dopamine D2 and D4 receptor mRNA in the adrenal gland, aldosterone-producing adenoma, and pheochromocytoma. J Clin Endocrinol Metab 2001;86: 4460–4467.
33 Chang H-W, Chu T-S, Huang H-Y, Chueh S-C, Wu V-C, Chen Y-M, Hsieh B-S, Wu K-D: Down-regulation of D2 dopamine receptor and increased protein kinase CW phosphorylation in aldosterone-producing adenoma play roles in aldosterone overproduction. J Clin Endocrinol Metab 2007;92:1863–1870.
34 Cartier D, Jégou S, Parmentier F, Lihrmann I, Louiset E, Kuhn J-M, Bastard C, Plouin P-F, Godin M, Vaudry H, Lefebvre H: Expression profile of serotonin4 (5-HT4) receptors in adrenocortical aldosterone-producing adenomas. Eur J Endocrinol 2005;153: 939–947.
35 Perraudin V, Delarue C, Lefebvre H, Do Rego J-L, Vaudry H, Kuhn J-M: Evidence for a role of vasopressin in the control of aldosterone secretion in primary aldosteronism: in vitro and in vivo studies. J Clin Endocrinol Metab 2006;91:1566–1572.
36 Spät A, Hunyady L: Control of aldosterone secretion: a model for convergence in cellular signaling pathways. Physiol Rev 2004;84:489–539.

37 Gummow BM, Winnay JN, Hammer GD: Convergence of Wnt signaling and steroidogenic factor-1 (SF-1) on transcription of the rat inhibin gene. J Biol Chem 2003;278:26572–26579.

38 Schinner S, Willenberg HS, Krause D, Schott M, Lamounier-Zepter V, Krug AW, Ehrhart-Bornstein M, Bornstein SR, Scherbaum WA: Adipocyte-derived products induce the transcription of the StAR promoter and stimulate aldosterone and cortisol secretion from adrenocortical cells through the Wnt signaling pathway. Int J Obes 2007;31:864–870.

39 Krug AW, Vleugels K, Schinner S, Lamounier-Zepter V, Ziegler CG, Bornstein SR, Ehrhart-Bornstein M: Human adipocytes induce an ERK1/2 MAP kinases-mediated upregulation of steroidogenic acute regulatory protein (StAR) and an angiotensin II-sensitization in human adrenocortical cells. Int J Obes 2007;31:1605–1616.

40 Inglis GC, Plouin PF, Friel EC, Davies E, Fraser R, Connell JMC: Polymorphic differences from normal in the aldosterone synthase gene (CYP11B2) in patients with primary hyperaldosteronism and adrenal tumour (Conn's syndrome). Clin Endocrinol 2001;54:725–730.

41 White PC, Slutsker L: Haplotype analysis of CYP11B2. Endocr Res 1995;21:437–442.

42 Müssig K, Wehrmann M, Horger M, Maser-Gluth C, Haring H-U, Overkamp D: Adrenocortical carcinoma producing 11-deoxycorticosterone: a rare cause of mineralocorticoid hypertension. J Endocrinol Invest 2005;28:61–65.

43 Griffing GT, Wilson TE, Holbrook MM, Dale SL, Jackson TK, Ullrich I, Melby JC: Plasma and urinary 19-nor-deoxycorticosterone in 17α-hydroxylase deficiency syndrome. J Clin Endocrinol Metab 1984;59:1011–1015.

44 Nicod J, Dick B, Frey FJ, Ferrari P: Mutation analysis of CYP11B1 and CYP11B2 in patients with increased 18-hydroxycortisol production. Mol Cell Endocrinol 2004;214:167–174.

45 Gallery EDM, Hunyor SN, Györy AZ: Plasma volume contraction: a significant factor in both pregnancy-associated hypertension (pre-eclampsia) and chronic hypertension in pregnancy. Q J Med 1979;48:593–602.

46 Spaanderman MEA, Aardenburg R, Ekhart THA, van Eyndhoven HWF, van der Heijden OWH, van Eyck J, de Leeuw PW, Peeters LLH: Non-pregnant circulatory volume status predicts subsequent pregnancy outcome in normotensive thrombophilic formerly preeclamptic women. Eur J Obstet Gynecol Reprod Biol 2001;95:218–221.

47 Gallery EDM, Brown MA: Control of sodium excretion in human pregnancy. Am J Kidney Dis 1987;9:290–295.

48 Mohaupt M: Molecular aspects of preeclampsia. Mol Aspects Med 2007;28:169–191.

49 Jensen E, Wood CE, Keller-Wood M: Chronic alterations in ovine maternal corticosteroid levels influence uterine blood flow and placental and fetal growth. Am J Physiol Regul Integr Comp Physiol 2005;288:R54–R61.

50 Robinson M: Salt in pregnancy. Lancet 1958;i:178–181.

51 Bower D: The influence of dietary salt intake on pre-eclampsia. J Obstet Gynaecol Br Commonw 1964;71:123–125.

52 Duley L, Henderson-Smart D, Meher S: Altered dietary salt for preventing pre-eclampsia, and its complications. Cochrane Database Syst Rev 2005;4:CD005548. DOI: 10.1002/14651858.

53 Salas SP, Marshall G, Gutiérrez BL, Rosso P: Time course of maternal plasma volume and hormonal changes in women with preeclampsia or fetal growth restriction. Hypertension 2006;47:203–208.

54 Shojaati K, Causevic M, Kadereit B, Dick B, Imobersteg J, Schneider H, Beinder E, Kashiwagi M, Frey BM, Frey FJ, Mohaupt MG: Evidence for compromised aldosterone synthase enzyme activity in preeclampsia. Kidney Int 2004;66:2322–2328.

55 Ehrlich EN, Nolten WE, Oparil S, Lindheimer MD: Mineralocorticoids in normal pregnancy. Perspect Nephrol Hypertens 1976;5:189–201.

56 Ulick S, Wang JZ, Morton DH: The biochemical phenotypes of two inborn errors in the biosynthesis of aldosterone. J Clin Endocrinol Metab 1992;74:1415–1420.

57 Zhang G, Rodriguez H, Fardella CE, Harris DA, Miller WL: Mutation T318M in the CYP11B2 gene encoding P450c11AS (aldosterone synthase) causes corticosterone methyl oxidase II deficiency. Am J Hum Genet 1995;57:1037–1043.

58 Mitsuuchi Y, Kawamoto T, Naiki Y, Miyahara K, Toda K, Kuribayashi I, Orii T, Yasuda K, Miura K, Nakao K, Imura H, Ulick S, Shizuta Y: Congenitally defective aldosterone biosynthesis in humans: the involvement of point mutations of the P-450_{C18} gene (CYP11B2) in CMO II deficient patients. Biochem Biophys Res Commun 1992;182:974–979.

59 Pascoe L, Curnow KM, Slutsker L, Rösler A, White PC: Mutations in the human *CYP11B2* (aldosterone synthase) gene causing corticosterone methyloxidase II deficiency. Proc Natl Acad Sci USA 1992;89:4996–5000.

60 Peter M, Bünger K, Drop SLS, Sippell WG: Molecular genetic study in two patients with congenital hypoaldosteronism (types I and II) in relation to previously published hormonal studies. Eur J Endocrinol 1998;139:96–100.

61 Fardella CE, Rodriguez H, Montero J, Zhang G, Vignolo P, Rojas A, Villarroel L, Miller WL: Genetic variation in P450c11AS in Chilean patients with low renin hypertension. J Clin Endocrinol Metab 1996;81:4347–4351.
62 Rösler A, White PC: Mutations in human 11β-hydroxylase genes: 11β-hydroxylase deficiency in Jews of Morocco and corticosterone methyl-oxidase II deficiency in Jews of Iran. J Steroid Biochem Mol Biol 1993;45:99–106.
63 Mitsuuchi Y, Kawamoto T, Miyahara K, Ulick S, Morton DH, Naiki Y, Kuribayashi I, Toda K, Hara T, Orii T, Yasuda K, Miura K, Yamamoto Y, Imura H, Shizuta Y: Congenitally defective aldosterone biosynthesis in humans: inactivation of the P-450_{C18} gene (CYP11B2) due to nucleotide deletion in CMO I deficient patients. Biochem Biophys Res Commun 1993;190:864–869.
64 Nomoto S, Massa G, Mitani F, Ishimura Y, Miyahara K, Toda K, Nagano I, Yamashiro T, Ogoshi S, Fukata J-I, Onishi S, Hashimoto K, Doi Y, Imura H, Shizuta Y: CMO I deficiency caused by a point mutation in exon 8 of the human CYP11B2 gene encoding steroid 18-hydroxylase ($P450_{C18}$). Biochem Biophys Res Commun 1997;234:382–385.
65 Peter M, Fawaz L, Drop SLS, Visser HKA, Sippell WG: Hereditary defect in biosynthesis of aldosterone: aldosterone synthase deficiency 1964–1997. J Clin Endocrinol Metab 1997;82:3525–3528.
66 Kayes-Wandover KM, Schindler REL, Taylor HC, White PC: Type 1 aldosterone synthase deficiency presenting in a middle-aged man. J Clin Endocrinol Metab 2001;86:1008–1012.
67 Ye P, Kenyon CJ, MacKenzie SM, Seckl JR, Fraser R, Connell JMC, Davies E: Regulation of aldosterone synthase gene expression in the rat adrenal gland and central nervous system by sodium and angiotensin II. Endocrinology 2003;144:3321–3328.
68 Brochu M, Lehoux J-G, Picard S: Effects of gestation on enzymes controlling aldosterone synthesis in the rat adrenal. Endocrinology 1997;138:2354–2358.
69 Brochu M, Roy-Clavel E, Picard S, St-Louis J: In vivo regulation of enzymes controlling aldosterone synthesis in pregnant rats. Endocr Res 1998;24:575–579.
70 Broughton-Pipkin F: Risk factors for preeclampsia. N Engl J Med 2001;344:925–926.
71 Pojoga L, Gautier S, Blanc H, Guyene TT, Poirier O, Cambien F, Benetos A: Genetic determination of plasma aldosterone levels in essential hypertension. Am J Hypertens 1998;11:856–860.
72 Maynard SE, Min J-Y, Merchan J, Lim K-H, Li J, Mondal S, Libermann TA, Morgan JP, Sellke FW, Stillman IE, Epstein FH, Sukhatme VP, Karumanchi SA: Excess placental soluble fms-like tyrosine kinase 1 (sFlt1) may contribute to endothelial dysfunction, hypertension, and proteinuria in preeclampsia. J Clin Invest 2003;111:649–658.
73 Nevo O, Soleymanlou N, Wu Y, Xu J, Kingdom J, Many A, Zamudio S, Caniggia I: Increased expression of sFlt-1 in in vivo and in vitro models of human placental hypoxia is mediated by HIF-1. Am J Physiol Regul Integr Comp Physiol 2006;291:R1085–R1093.
74 Farese S, Shojaati K, Kadereit B, Frey FJ, Mohaupt MG: Blood pressure reduction in pregnancy by sodium chloride. Nephrol Dial Transplant 2006;21:1984–1987.

Markus G. Mohaupt, MD
Division of Hypertension of the University of Bern
Department of Nephrology/Hypertension, University Hospital Bern
CH–3010 Bern (Switzerland)
Tel. +41 31 632 9731, Fax +41 31 632 9734, E-Mail markus.mohaupt@insel.ch

Flück CE, Miller WL (eds): Disorders of the Human Adrenal Cortex.
Endocr Dev. Basel, Karger, 2008, vol 13, pp 145–158

Fetal Programming of Adrenal Androgen Excess: Lessons from a Nonhuman Primate Model of Polycystic Ovary Syndrome

David H. Abbott[a–c] · Rao Zhou[a] · Ian M. Bird[a,c] · Daniel A. Dumesic[b,d] · Alan J. Conley[e]

[a]Department of Obstetrics and Gynecology, [b]Wisconsin National Primate Research Center and [c]Endocrinology-Reproductive Physiology Training Program, University of Wisconsin, Madison, Wisc., [d]Reproductive Medicine and Infertility Associates, Woodbury, Minn., and [e]Population Health and Reproduction, School of Veterinary Medicine, University of California-Davis, Davis, Calif., USA

Abstract

Adrenal androgen excess is found in adult female rhesus monkeys previously exposed to androgen treatment during early gestation. In adulthood, such prenatally androgenized female monkeys exhibit elevated basal circulating levels of dehydroepiandrosterone sulfate (DHEAS), typical of polycystic ovary syndrome (PCOS) women with adrenal androgen excess. Further androgen and glucocorticoid abnormalities in PA female monkeys are revealed by acute ACTH stimulation: DHEA, androstenedione and corticosterone responses are all elevated compared to responses in controls. Pioglitazone treatment, however, diminishes circulating DHEAS responses to ACTH in both prenatally androgenized and control female monkeys, while increasing the 17-hydroxyprogesterone response and reducing the DHEA to 17-hydroxyprogesterone ratio. Since 60-min post-ACTH serum values for 17-hydroxyprogesterone correlate negatively with basal serum insulin levels (all female monkeys on pioglitazone and placebo treatment combined), while similar DHEAS values correlate positively with basal serum insulin levels, circulating insulin levels may preferentially support adrenal androgen biosynthesis in both prenatally androgenized and control female rhesus monkeys. Overall, our findings suggest that differentiation of the monkey adrenal cortex in a hyperandrogenic fetal environment may permanently upregulate adult adrenal androgen biosynthesis through specific elevation of 17,20-lyase activity in the zona fasciculata-reticularis. As adult prenatally androgenized female rhesus monkeys closely emulate PCOS-like symptoms, excess fetal androgen programming may contribute to adult adrenal androgen excess in women with PCOS.

In contrast to many mammals, the primate adrenal cortex is well endowed with androgen biosynthetic machinery located within two developmentally specific adrenocortical zones: the fetal zone and the zona reticularis [1, 2]. While the androgen-producing fetal zone develops during gestation and regresses shortly after birth, the androgenic zona reticularis, the innermost cortical zone, develops later in life and functionally matures at adrenarche [3–5]. Both adrenocortical zones produce the adrenal androgen dehydroepiandrosterone (DHEA) and its sulfoconjugate DHEAS, while androstenedione is produced in relatively small amounts due to the reduced expression of 3β-hydroxysteroid dehydrogenase II (3β-HSD II) [2].

The androgenic predilection of the zona reticularis stems from the preferential conversion of 17-hydroxypregnenolone to DHEA (with onward sulfoconjugation to DHEAS), rather than to 17-hydroxyprogesterone (and onward conversion to cortisol), because of (1) diminished expression of 3β-HSD II, together with relatively high expression of (2) the androgenic biosynthetic enzyme P450c17 (17α-hydroxylase/17,20-lyase), (3) the catalytic accessory protein, cytochrome b5, required for efficient 17,20-lyase activity, and (4) an obligate electron donor for P450c17 activity, NADPH cytochrome P450 oxidoreductase [1, 2] (fig. 1). Implicit in this steroidogenic activity is the fact that primates, in contrast to rodents, derive most of their androgen biosynthesis via the Δ5 steroidogenic pathway (fig. 1) because 17-hydroxyprogesterone, as an inefficient substrate for primate 17,20-lyase, is negligibly converted to androstenedione [1, 2].

Old World monkeys [i.e., rhesus macaques *(Macaca mulatta)* from the Indian subcontinent], great apes and humans exhibit analogous maturation of the androgenic adrenal zona reticularis [6], with all undergoing adrenarche, or increased adrenal DHEA and DHEAS secretion [2] coincident with greatly reduced expression of 3β-HSD II within the zona reticularis. Monkeys differ from great apes and humans, however, with respect to the developmental timing of adrenarche. In contrast to progressive differentiation of the zona reticularis from neonatal to adolescent stages in great apes and humans [2, 7], such progressive maturation of the zona reticularis in rhesus monkeys and baboons occurs during late fetal to neonatal life coincident with regression of the fetal zone [2, 5]. Rhesus monkeys, therefore, have close similarities to adrenal androgen physiology found in humans making them appropriate animal models for investigating pathophysiological mechanisms underlying adrenal androgen excess.

This chapter focuses on a potential fetal programming etiology for adrenal androgen excess in polycystic ovary syndrome (PCOS), a hyperandrogenic, infertility and metabolic disorder found in 6–7% of reproductive-aged women [8–10]. PCOS accounts for 82% of women presenting with androgen excess [11] and between 20 and 30% of PCOS women exhibit adrenal androgen excess, manifest by elevated circulating levels of DHEA, DHEAS and androstenedione [12–15], as well as specific DHEA and androstenedione hyperandrogenic responses to ACTH [16]. Moreover, DHEAS levels are generally elevated in most women with PCOS, suggesting a prevalent adrenal contribution to this hyperandrogenic condition because of the unique adrenal origins of the conjugated steroid [14]. While hyperandrogenic adrenal responses to ACTH in PCOS women are not necessarily accompanied by other

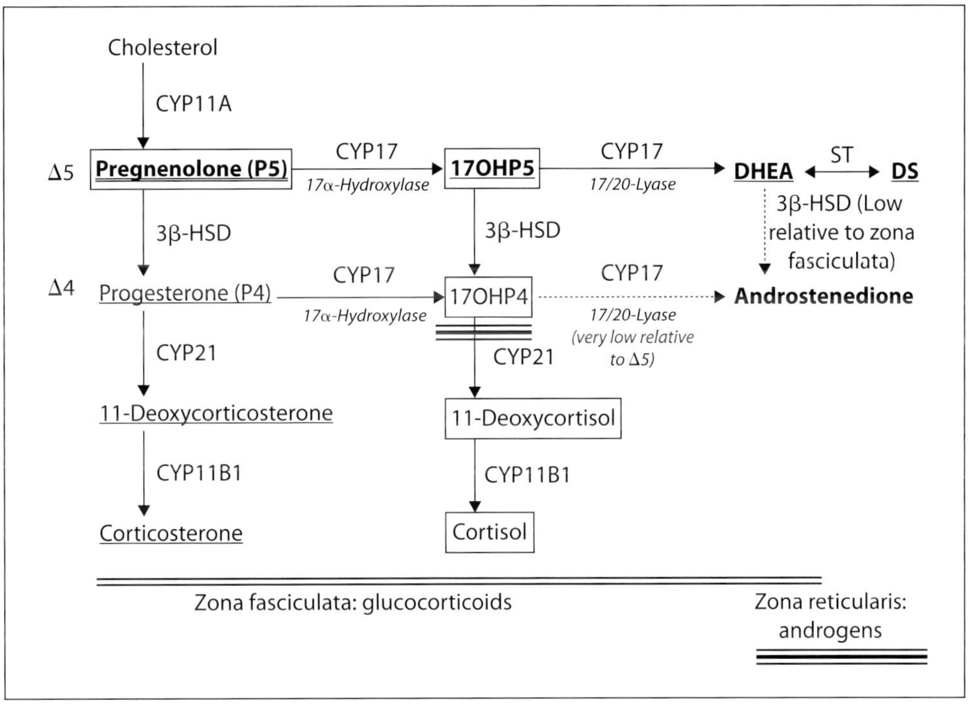

Fig. 1. Steroid biosynthesis in the mid to inner adrenal cortical zones, zona fasciculata and zona reticularis, of Old World primates and humans. Steroids in bold represent the predominant pathway for androgen biosynthesis, steroids within boxes represent the predominant pathway for cortisol biosynthesis, and underlined steroids represent the predominant pathway for corticosterone biosynthesis. The long arrows reflect the proposed enhanced 3β-HSD II and 17,20-lyase enzymatic function in PA female monkeys, while the dotted arrows reflect relative low enzymatic activity in both control and PA females. Double underlining reflects the extent of zona fasciculata enzymatic activity, while triple underlining reflects the extent of zona reticularis enzymatic activity. 17OHP5 = 17α-Hydroxypregnenolone; 17OHP4 = 17α-hydroxyprogesterone; CYP11A = P450scc; CYP17 = P450c17; ST = SULT2A1 sulfotransferase; DS = dehydroepiandrosterone sulfate; CYP21 = P450c21; CYP11B1 = P450c11 [modified from 31].

remarkable adrenal abnormalities of the zona glomerulosa and zona fasciculata, or the hypothalamic-pituitary-adrenal axis [16, 17], some PCOS studies suggest a generalized hypersecretion of adrenal steroids [15], or ACTH hyperresponsiveness to corticotropin-releasing hormone (CRH) [18, 19]. The latter, however, may reflect diminished somatostatin release in PCOS women [20], particularly since somatostatin treatment reduces CRH-mediated ACTH release in PCOS women [21].

To explore a fetal programming origin for adrenal androgen excess in PCOS, adrenal androgen abnormalities were examined in adult female rhesus monkeys exposed in utero to androgen excess. These prenatally androgenized female rhesus monkeys closely mimic the reproductive, infertility and metabolic phenotypes found in PCOS women [22] and show a similar degree of adrenal androgen excess.

Prenatally Androgenized Female Rhesus Monkeys as Models of Adrenal Androgen Excess

Prenatally androgenized female rhesus monkeys are generated by exposure to fetal male levels of testosterone during early or late gestation. Their pregnant mothers receive subcutaneous injections of 10–15 mg testosterone propionate for up to 41 consecutive days, starting on gestation days 40–44 (early) or 100/110 (late) in approximately 165-day term pregnancies [23]. Such experimentally induced androgen excess generates PCOS-like phenotypes when prenatally androgenized females reach adulthood [22]. Early gestation-exposed prenatally androgenized female monkeys show irregular or absent ovulatory menstrual cycles, ovarian hyperandrogenism, enlarged polyfollicular ovaries and luteinizing hormone hypersecretion, in addition to insulin resistance, diminished insulin secretion, increased incidence of type 2 diabetes, visceral adiposity and hyperlipidemia [22, 24–27]. Female monkeys similarly exposed to androgen excess during late gestation also exhibit an adult PCOS-like phenotype, but without obvious abnormalities in luteinizing hormone and insulin secretion or in insulin action [22].

Both types of adult prenatally androgenized monkeys demonstrate ovarian androgen excess. For example, 24 h after an intramuscular injection of 200 IU of recombinant human chorionic gonadotropin, both early and late gestation-exposed prenatally androgenized female monkey groups exhibit elevated circulating testosterone levels compared to controls [22]. Early gestation-exposed prenatally androgenized monkeys also show an elevated ovarian 17-hydroxyprogesterone response to recombinant human chorionic gonadotropin [28] together with elevated basal testosterone levels [27, 29]. These females with more obvious signs of fetal programming of adult hyperandrogenism were thus studied to determine whether they had adrenal androgen excess comparable to that found in PCOS women.

Potential Adrenal Hyperandrogenism in Infant Prenatally Androgenized Monkeys Exposed to Fetal Androgen Excess during Early Gestation

In rhesus monkeys, early infancy coincides not only with regression of the androgen-producing fetal zone of the adrenal, but also with the simultaneous maturation of the androgenic zona reticularis [2, 5]. From at least birth to 1 month of age, prenatally androgenized female monkeys exhibit elevated circulating levels of androstenedione [23], indicative of adrenal and/or ovarian androgen excess. Our preliminary findings suggest that adrenal androgen excess does contribute to this hyperandrogenism in infant prenatally androgenized females, given the elevated protein expression of adrenal P450 oxidoreductase and possibly cytochrome b5 [Conley and Abbott, unpubl. results]. Interestingly, infant male rhesus monkeys exhibit an adrenal-dependent rise in circulating testosterone levels between 4 and 25 weeks of age [30] so that exposure to fetal androgen excess (fetal male or prenatally androgenized female) during early gestation may permanently enhance adrenal androgen biosynthesis following birth.

Adrenal Hyperandrogenism in Adult Prenatally Androgenized Monkeys Exposed to Fetal Androgen Excess during Early Gestation

As a sign of functional adrenal hyperandrogenism, prenatally androgenized female monkeys exposed to androgen excess during early gestation exhibit elevated basal circulating levels of DHEA [31] and DHEAS [28]. Elevated basal DHEAS levels are a primary indicator of adrenal androgen excess in women with PCOS [15, 32, 33] suggesting that prenatally androgenized monkeys may closely emulate their PCOS human counterparts in this adrenal abnormality. To examine prenatally androgenized monkey adrenal steroid dysfunction in more detail, we utilize acute ACTH stimulation (at 07:30 to 08:00 h) administered 14.5–16 h after dexamethasone treatment (Dex-ACTH) on days 2–6 of the menstrual cycle or during a 30-day anovulatory interval. We employ a high intravenous dose of ACTH (50 μg; ~5.5 μg/kg of human $ACTH_{1-39}$) to saturate the preferred P450c17 Δ5 pathway and to reveal any minor shifts in steroidogenesis from 17-hydroxypregnenolone to progesterone via increased 3β-HSD II activity [31] (fig. 1). Such ACTH-mediated adrenocortical stimulation further illustrates the functional adrenal hyperandrogenism of prenatally androgenized female monkeys, with elevated serum DHEA and androstenedione responses to ACTH (fig. 2) [31] accompanied by normal serum 17-hydroxyprogesterone and cortisol responses to ACTH. Note that the latter result suggests an additional, but normal, zona reticularis origin of circulating 17-hydroxyprogesterone, since P450c21 in the zona fasciculata would rapidly metabolize this product to deoxycortisol (fig. 1).

In these prenatally androgenized female monkeys, the relatively greater increase in serum DHEA compared to serum 17-hydroxyprogesterone, androstenedione and DHEAS in response to Dex-ACTH causes (1) an increased ratio of serum DHEA to 17-hydroxyprogesterone and (2) diminished ratios of serum DHEAS to DHEA, and serum androstenedione to DHEA [31]. Such adrenal androgen excess in early gestation-exposed prenatally androgenized female monkeys corresponds with selectively increased 17,20-lyase activity in the zona fasciculata (elevated DHEA and androstenedione levels) as well as in the zona reticularis (elevated DHEA levels). This enhanced efficiency of 17,20-lyase activity in the prenatally androgenized monkey adrenal, however, occurs without enhanced 17α-hydroxylase activity, since basal and ACTH-induced levels of 17-hydroxyprogesterone (from the zona reticularis) and cortisol (from the zona fasciculata) are normal. A putative molecular basis for an isolated lyase activity increase in prenatally androgenized monkey adrenals could be increased serine phosphorylation of P450c17 [31] possibly facilitated by an increase of cytochrome

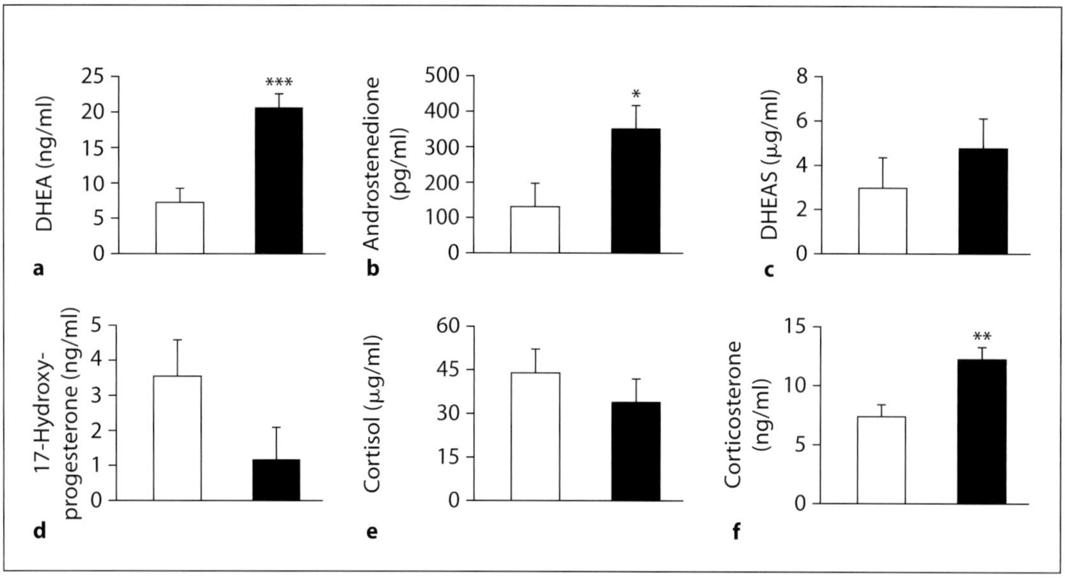

Fig. 2. The differential in circulating adrenal steroid values between 0 and 60 min following an intravenous injection of 50 µg ACTH in adult control (□) and prenatally androgenized (■) female rhesus monkeys for adrenal androgen-related steroids: DHEA (**a**), androstenedione (**b**) and DHEAS (**c**), as well as adrenal glucocortoicoid-related steroids: 17-hydroxyprogesterone (**d**), cortisol (**e**) and corticosterone (**f**) [modified from 31]. $*p > 0.05$; $**p > 0.03$; $***p > 0.01$.

b5 or P450 oxidoreductase, or most likely in view of the involvement of insulin sensitivity, a combination of all three. Additionally, conversion of excessive DHEA to DHEAS via SULT2A1 sulfotransferase in the zona reticularis, and to androstenedione via 3β-HSD II in the zona fasciculata (and to a greatly reduced extent in the zona reticularis; fig. 1), could provide one explanation for the elevated levels of these latter two androgenic adrenal steroids in the absence of an altered cortisol response.

To account for the increased corticosterone response to ACTH combined with adrenal hyperandrogenism in prenatally androgenized female monkeys, however, additional enzymatic changes in adrenal steroid biosynthesis and metabolism must exist beyond enhanced 17,20-lyase in the zona fasciculata and zona reticularis. While pregnenolone is the preferred substrate for cortisol biosynthesis in primates, a small amount of progesterone byproduct is the basis for corticosterone biosynthesis by the zona fasciculata (fig. 1) [2, 6, 34, 35]. The corticosterone response is produced primarily in the primate zona fasciculata [2, 6], and is elevated following ACTH stimulation in prenatally androgenized females, so we must assume 3β-HSD II activity is elevated in the zona fasciculata of these monkeys. This hypothesis agrees with a 3–4% decrease in the efficiency of cortisol biosynthesis in the zona fasciculata (derived from the increased ratio of corticosterone to cortisol at 60 min after ACTH administration)

Table 1. Differences in mean ± SEM circulating adrenal steroid values between 0 and 60 min following an intravenous injection of 50 μg ACTH in the combined group of control and prenatally androgenized female rhesus monkeys receiving placebo or pioglitazone treatment [27]

Steroid	0- to 60-min placebo	0- to 60-min pioglitazone	p value
17α-Hydroxyprogesterone, ng/ml	2.3 ± 1.0	3.6 ± 1.0	NS
Cortisol, μg/dl	31.9 ± 3.8	36.7 ± 3.9	NS
DHEA, ng/ml	16.7 ± 1.7	17.9 ± 1.8	NS[1]
DHEAS, μg/dl	6.9 ± 1.5	2.2 ± 1.6	<0.05
Androstenedione, pg/ml	213 ± 52	345 ± 78	NS
Corticosterone, ng/ml	11.1 ± 1.2	10.8 ± 1.1	NS

NS = $p > 0.05$.
[1]The difference in serum DHEA levels between 0 and 60 min post-ACTH injection for prenatally androgenized female monkeys exceeds that in control females, regardless of placebo or pioglitazone treatment (treatment values combined: prenatally androgenized, 20.0 ± 1.8 ng/ml; controls, 14.5 ± 1.8 ng/ml; $p < 0.05$).

[34] of prenatally androgenized female monkeys, as well as an ~14-fold increase in corticosterone versus an ~8-fold increase in cortisol following ACTH administration [31]. While enhanced 3β-HSD II activity exists in some PCOS women with adrenal androgen excess [36], it is unclear whether this phenomenon translates into corticosterone hyperresponsiveness to ACTH in humans.

Insulin Is Associated with Adrenal Androgen Excess in Prenatally Androgenized Female Monkeys

As in PCOS [37], insulin is associated with the pathophysiological mechanism of reproductive and endocrine abnormalities in prenatally androgenized female rhesus monkeys, as evidenced by a 6-month study of pioglitazone treatment, a thiazolidinedione-based insulin sensitizer. Pioglitazone treatment not only diminishes insulin resistance and reduces circulating insulin levels, but also normalizes menstrual cyclicity and diminishes ovarian hyperandrogenism in most prenatally androgenized female monkeys [27], similar to its effects in PCOS women [38, 39].

Similar to its effects in PCOS women [40, 41], pioglitazone treatment of prenatally androgenized female monkeys diminishes aspects of adrenal androgen excess. Serum DHEAS responses to ACTH are reduced in both prenatally androgenized and control female monkeys (table 1), analogous to findings of reduced basal serum DHEAS levels in PCOS women undergoing treatment with troglitazone [42]. In prenatally androgenized female monkeys alone, pioglitazone normalizes ($p < 0.02$) the serum DHEA to 17-hydroxyprogesterone ratio at 60 min following ACTH administration (prenatally androgenized, placebo: 22.0 ng/ml (16.9, 28.7), pioglitazone: 8.1 (6.1, 10.8); control, placebo: 7.3 (1.9, 28.8), pioglitazone: 7.7 (2.0, 26.7) (backtransformed logarithmic mean; 95% confidence interval). Serum DHEA responses to ACTH, however, remain elevated at their previously reported values (compare table 1 to fig. 2) in prenatally androgenized compared to control female monkeys regardless of whether they receive pioglitazone or placebo treatment.

Together, these pioglitazone-mediated changes in adrenal steroidogenesis predominantly implicate a specific reduction in 17,20-lyase activity following the pioglitazone-mediated decline in circulating insulin levels. Such a notion is consistent with the negative correlation between 60-min post-ACTH 17-hydroxyprogesterone levels and basal insulin (fig. 3a), the positive correlation between 60-min post-ACTH DHEAS levels and basal insulin (fig. 3b) and the overall negative correlation between 0-min pre-ACTH 17-hydroxyprogesterone and DHEAS levels (fig. 3c) in all female monkeys combined. Decreased serine phosphorylation of P450c17 [33, 43], decreased cytochrome b5 activity [43, 44], decreased concentrations of the electron-donating redox partner for 17,20-lyase, P450 oxidoreductase [45, 46], or a combination of all three may provide a molecular basis for the relatively diminished 17,20-lyase activity observed in pioglitazone-treated prenatally androgenized female monkeys. It is unlikely that such diminished adrenal androgen excess is a direct result of pioglitazone action. Pioglitazone, at therapeutic doses, increases P450 oxidoreductase expression [47, 48], increases melanocortin 2 (ACTH) receptor expression and cortisol responses to ACTH [49], and downregulates expression of P450c17 and 3β-HSD II [50].

The finding of pioglitazone-mediated decrease in 17,20-lyase activity agrees with compensatory hyperinsulinemia from insulin resistance playing a causal role in adrenal hyperandrogenism in PCOS [51, 52], and with insulin-enhanced adrenal androgen responses to ACTH in PCOS [53]. Moreover, if enhanced 17,20-lyase activity extends into the adrenal zona fasciculata, it can also contribute to proportionately increased DHEA biosynthesis, perhaps even at the expense of 17α-hydroxyprogesterone and cortisol production, as found in the highly inbred Angora goat [54]. Such an expansive function for 17,20-lyase in the zona fasciculata might also contribute towards the lack of concomitant increase in 17α-hydroxyprogesterone and cortisol (fig. 1) following ACTH in prenatally androgenized versus control female monkeys.

Insulin stimulation of P450c17 in the adrenal cortex of PCOS women, if similar to its action in theca cells from polycystic ovaries, acts through the PI3-kinase branch of the insulin signaling pathway, to synergize with additional coactivation by cAMP/kinase A [55]. Such insulin overstimulation of P450c17 under circumstances of insulin resistance at the level of glucose metabolism would suggest that steroidogenesis and glucose metabolism are regulated by alternate signaling events distal to

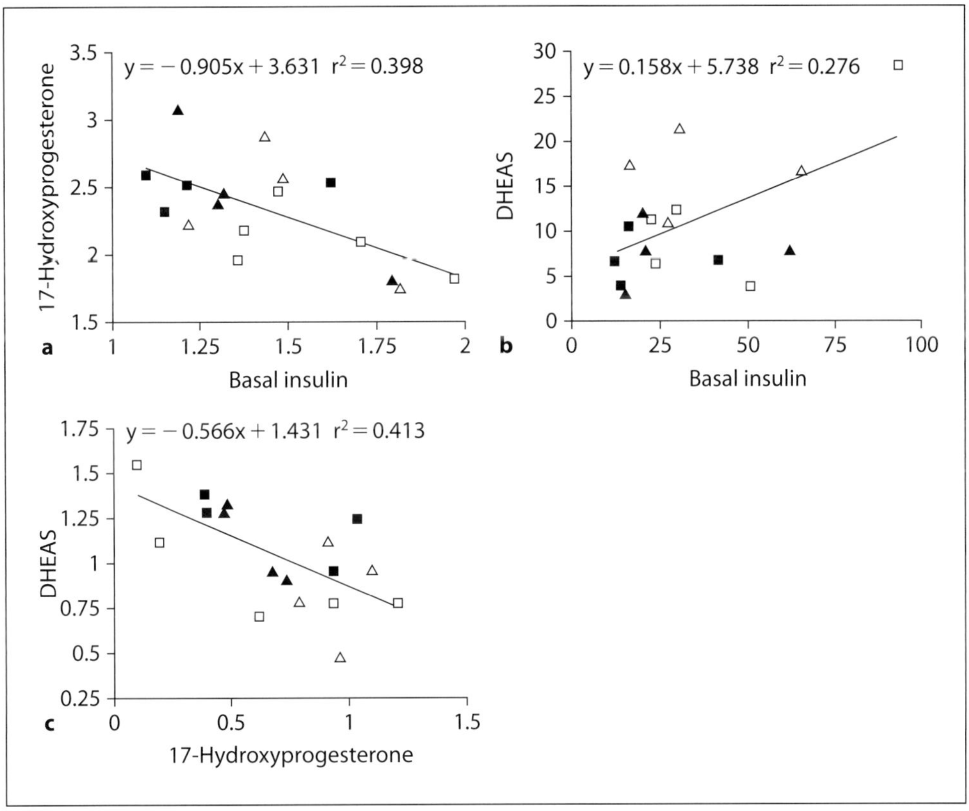

Fig. 3. Associations ($p < 0.05$) between serum values of 17-hydroxyprogesterone at 60 min following an intravenous injection of 50 μg ACTH and basal insulin (**a**), DHEAS at 60 min following an intravenous injection of 50 μg ACTH and basal insulin (**b**), and DHEAS and 17-hydroxyprogesterone at 0 min following an intravenous injection of 50 μg ACTH (**c**) in control (△ = placebo treatment [27]; ▲ = pioglitazone treatment [27]) and PA (□ = placebo; ■ = pioglitazone) female monkeys. Data are expressed as log_{10}-transformed serum values only in **a** and **c**. Units used: 17-hydroxyprogesterone, log_{10} ng/ml; insulin, log_{10} μU/ml (**a**); DHEAS, μg/dl; insulin, μU/ml (**b**), and DHEAS, log_{10} μg/dl; 17-hydroxyprogesterone, log_{10} ng/ml (**c**).

PI3-kinase activation [55], and would explain the absence of insulin resistance in terms of steroidogenic enzyme regulation in the ovary [56] as well as the adrenal.

Potential Fetal Programming Origins of Adrenal Androgen Excess

In the adrenal cortex, androgens are implicated in the regulation of their own biosynthesis. For example, adrenocortical cells express androgen receptors [57, 58], flutamide (an androgen receptor antagonist) diminishes adrenal androgen biosynthesis in PCOS women [59, 60], while testosterone can directly affect adrenal androgen biosynthesis [61]. Taken together, a hyperandrogenic environment during fetal life may influence adrenal cortical differentiation in a manner that enhances fetal zone androgen biosynthesis.

Alternatively, insulin appears to upregulate androgen biosynthesis in the adrenal cortex of normal [61] and PCOS women [53], and in the polycystic ovary [36]. Since both fetal [29] and infant [62] prenatally androgenized female rhesus monkeys exhibit enhanced insulin secretion, the fetal zone and infant zona reticularis may be susceptible to increased insulin-mediated serine phosphorylation of P450c17, thus preferentially enhancing 17,20-lyase activity by increasing the affinity of P450c17 for its electron donor, P450 oxidoreductase [33, 43]. Therefore, fetal androgen excess may induce relative insulin hypersecretion in exposed female fetuses and infants, which in turn programs adrenal hyperandrogenism.

Conclusions and Therapeutic Considerations

Androgen excess, during early to midgestation in female rhesus monkeys, causes a remarkably close phenocopy of adrenal hyperandrogenism found in 20–30% of PCOS women. The ability to experimentally induce adrenal androgenism by fetal programming further suggests that specific changes in adrenal gene expression, perhaps from epigenetic alterations of gene methylation [63, 64], may contribute to adrenal androgen excess beyond genotypic determination [65]. While the molecular mechanisms underlying adrenal pathophysiology in PCOS women and prenatally androgenized female monkeys are incompletely understood, isolated enhancement of adrenocortical 17,20-lyase activity appears to be a common attribute, as evidenced by the elevated expression of P450 oxidoreductase, and possibly cytochrome b5, found in the adrenal cortex of our prenatally androgenized female monkeys [Conley and Abbott, unpubl. results]. It remains to be determined whether such molecular underpinnings of adrenal androgen excess are induced by the direct actions of testosterone, its androgenic or estrogenic metabolites, or by the hyperandrogenic induction of fetal and infant insulin hypersecretion. Regardless, our findings of fetal programming of adrenal androgen excess in a nonhuman primate model for PCOS suggest that such hyperandrogenism may be a consequence of fetal environmental influences rather than a primary, intrinsic defect.

Therapeutic use of low-dose flutamide reliably diminishes adrenal androgen excess in PCOS women [66]. Unfortunately, the use of flutamide to investigate possible direct effects of androgen excess in the fetal female environment on adrenal androgen biosynthesis is difficult due to its permanent effects on somatosensory and cognitive function in rhesus monkeys exposed during fetal development [67, 68]. On the other hand, insulin sensitizers alone (PCOS women [40, 41]; prenatally androgenized monkeys [27]) or in combination with flutamide [69], also ameliorate adrenal androgen excess in both PCOS women and adult female prenatally androgenized monkeys, without inducing hepatoxicity. Therefore, nonteratogenic insulin sensitizers such as metformin [70], in addition to their potential benefits in reducing pregnancy complications in PCOS [71], may prove useful in exploring how androgen-induced fetal hyperinsulinism programs adrenal hyperandrogenism. Future studies are required to determine whether insulin sensitizer treatment of pregnant PCOS women or pregnant testosterone propionate-treated monkeys will prevent postnatal adrenal hyperandrogenism and other phenotypic characteristics of PCOS in their daughters and confirm a specific programming mechanism.

Acknowledgments

This work was supported by NIH grants R01 RR013635, P50 HD044405, U01 HD044650 and P51 RR000167. This research was conducted at a facility constructed with support from Research Facilities Improvement Program grant No. RR15459 and RR020141.

References

1 Conley AJ, Bird IM: The role of cytochrome P450 17 alpha-hydroxylase and 3 beta-hydroxysteroid dehydrogenase in the integration of gonadal and adrenal steroidogenesis via the delta 5 and delta 4 pathways of steroidogenesis in mammals. Biol Reprod 1997; 56:789–799.

2 Conley AJ, Pattison JC, Bird IM: Variations in adrenal androgen production among (nonhuman) primates. Semin Reprod Med 2004;22:311–326.

3 Mesiano S, Jaffe RB: Developmental and functional biology of the primate fetal adrenal cortex. Endocr Rev 1997;18:378–403.

4 Mapes S, Tarantal AF, Parker CR, Moran FM, Bahr JM, Pyter L, Conley AJ: Adrenocortical cytochrome b5 expression during fetal development of the rhesus macaque. Endocrinology 2002;143:1451–1458.

5 Nguyen AD, Conley AJ: Adrenal androgens in humans and nonhuman primates: Production, zonation and regulation; in Flueck CE, Miller WL (eds): Disorders of the Human Adrenal Cortex. Basel, Karger, 2008, pp 33–54.

6 Arlt W, Martens JW, Song M, Wang JT, Auchus RJ, Miller WL: Molecular evolution of adrenarche: structural and functional analysis of P450c17 from four primate species. Endocrinology 2002;143:4665–4672.

7 Sklar CA, Kaplan SL, Grumbach MM: Evidence for dissociation between adrenarche and gonadarche: studies in patients with idiopathic precocious puberty, gonadal dysgenesis, isolated gonadotropin deficiency, and constitutionally delayed growth and adolescence. J Clin Endocrinol Metab 1980;51:548–556.

8 Diamanti-Kandarakis E, Kouli CR, Bergiele AT, Filandra FA, Tsianateli TC, Spina GG, Zapanti ED, Bartzis MI: A survey of the polycystic ovary syndrome in the Greek island of Lesbos: hormonal and metabolic profile. J Clin Endocrinol Metab 1999;84: 4006–4011.

9 Asuncion M, Calvo RM, San Millan JL, Sancho J, Avila S, Escobar-Morreale HF: A prospective study of the prevalence of the polycystic ovary syndrome in unselected Caucasian women from Spain. J Clin Endocrinol Metab 2000;85:2434–2438.

10 Azziz R, Woods KS, Reyna R, Key TJ, Knochenhauer ES, Yildiz BO: The prevalence and features of the polycystic ovary syndrome in an unselected population. J Clin Endocrinol Metab 2004;89:2745–2749.

11 Azziz R, Sanchez LA, Knochenhauer ES, Moran C, Lazenby J, Stephens KC, Taylor K, Boots LR: Androgen excess in women: experience with over 1000 consecutive patients. J Clin Endocrinol Metab 2004; 89:453–462.

12 Wild RA, Umstot ES, Andersen RN, Ranney GB, Givens JR: Androgen parameters and their correlation with body weight in one hundred thirty-eight women thought to have hyperandrogenism. Am J Obstet Gynecol 1983;146:602–606.

13 Carmina E, Koyama T, Chang L, Stanczyk FZ, Lobo RA: Does ethnicity influence the prevalence of adrenal hyperandrogenism and insulin resistance in polycystic ovary syndrome? Am J Obstet Gynecol 1992;167:1807–1812.

14 Kumar A, Woods KS, Bartolucci AA, Azziz R: Prevalence of adrenal androgen excess in patients with the polycystic ovary syndrome (PCOS). Clin Endocrinol (Oxf) 2005;62:644–649.
15 Yildiz BO, Azziz R: The adrenal and polycystic ovary syndrome. Rev Endocr Metab Disord 2007;8: 331–342.
16 Azziz R, Black V, Hines GA, Fox LM, Boots LR: Adrenal androgen excess in the polycystic ovary syndrome: sensitivity and responsivity of the hypothalamic-pituitary-adrenal axis. J Clin Endocrinol Metab 1998;83:2317–2323.
17 Pasquali R, Patton L, Pocognoli P, Cognigni GE, Gambineri A: 17-Hydroxyprogesterone responses to gonadotropin-releasing hormone disclose distinct phenotypes of functional ovarian hyperandrogenism and polycystic ovary syndrome. J Clin Endocrinol Metab 2007;92:4208–4217.
18 Kondoh Y, Uemura T, Ishikawa M, Yokoi N, Hirahara F: Classification of polycystic ovary syndrome into three types according to response to human corticotropin-releasing hormone. Fertil Steril 1999;72: 15–20.
19 Lanzone A, Petraglia F, Fulghesu AM, Ciampelli M, Caruso A, Mancuso S: Corticotropin-releasing hormone induces an exaggerated response of adrenocorticotropic hormone and cortisol in polycystic ovary syndrome. Fertil Steril 1995;63:1195–1199.
20 Wu XK, Wang CH, Su YH: Responses of somatostatin, beta-endorphin and dynorphin A to a glucose load in two groups of women with polycystic ovarian syndrome. Horm Res 1996;46:59–63.
21 Lanzone A, Fulghesu AM, Guido M, Cucinelli F, Caruso A, Mancuso S: Somatostatin treatment reduces the exaggerated response of adrenocorticotropin hormone and cortisol to corticotropin-releasing hormone in polycystic ovary syndrome. Fertil Steril 1997;67:34–39.
22 Abbott DH, Barnett DK, Bruns CM, Dumesic DA: Androgen excess fetal programming of female reproduction: a developmental aetiology for polycystic ovary syndrome? Hum Reprod Update 2005;11: 357–374.
23 Abbott DH, Barnett DK, Turk JM, Levine JE, Padmanabhan V, Dumesic DA, Jacoris S, Tarantal AF: Endocrine antecedents of polycystic ovary syndrome in fetal and infant prenatally androgenized female rhesus monkeys. Biol Reprod 2008; Apr 2; [Epub ahead of print].
24 Dumesic DA, Schramm RD, Abbott DH: Early origins of polycystic ovary syndrome. Reprod Fertil Dev 2005;17:349–360.
25 Eisner JR, Dumesic DA, Kemnitz JW, Abbott DH: Timing of prenatal androgen excess determines differential impairments in insulin secretion and action in adult female rhesus monkeys. J Clin Endocrinol Metab 2000;85:1206–1210.
26 Bruns CM, Baum ST, Colman RJ, Dumesic DA, Eisner JR, Jensen MD, Whigham LD, Abbott DH: Prenatal androgen excess negatively impacts body fat distribution in a nonhuman primate model of polycystic ovary syndrome. Int J Obes (Lond) 2007;31: 1579–1585.
27 Zhou R, Bruns CM, Bird IM, Kemnitz JW, Goodfriend TL, Dumesic DA, Abbott DH: Pioglitazone improves insulin action and normalizes menstrual cycles in a majority of prenatally androgenized female rhesus monkeys. Reprod Toxicol 2007;23: 438–448.
28 Eisner JR, Barnett MA, Dumesic DA, Abbott DH: Ovarian hyperandrogenism in adult female rhesus monkeys exposed to prenatal androgen excess. Fertil Steril 2002;77:167–172.
29 Abbott DH, Barnett DK, Bruns CM, Dunaif A, Dumesic DA: Transient hyperglycemia in both mother and fetus from experimental induction of maternal androgen excess in a nonhuman primate model for polycystic ovary syndrome (abstract P3-113). 88th Annual Meeting of the Endocrine Society, Boston, 2006.
30 Plant TM, Zorub DS: A study of the role of the adrenal glands in the initiation of the hiatus in gonadotropin secretion during prepubertal development in the male rhesus monkey *(Macaca mulatta)*. Endocrinology 1984;114:560–565.
31 Zhou R, Bird IM, Dumesic DA, Abbott DH: Adrenal hyperandrogenism is induced by fetal androgen excess in a rhesus monkey model of polycystic ovary syndrome. J Clin Endocrinol Metab 2005; 90:6630–6637.
32 Carmina E, Lobo RA: Prevalence and metabolic characteristics of adrenal androgen excess in hyperandrogenic women with different phenotypes. J Endocrinol Invest 2007;30:111–116.
33 Zhang LH, Rodriguez H, Ohno S, Miller WL: Serine phosphorylation of human P450c17 increases 17, 20-lyase activity: implications for adrenarche and the polycystic ovary syndrome. Proc Natl Acad Sci USA 1995;92:10619–10623.
34 Pattison JC, Abbott DH, Saltzman W, Mapes SM, Moran FM, Corbin CJ, Nguyen AD, Pryce CR, Allen AJ, Conley AJ, Bird IM: Male marmoset monkeys express an adrenal fetal zone at birth but not a zona reticularis in adulthood. Endocrinology 2005; 146:365–374.

35 Pattison JC, Saltzman W, Abbott DH, Hogan BK, Nguyen AD, Husen B, Einspanier A, Conley AJ, Bird IM: Gender and gonadal status differences in zona reticularis expression in marmoset monkey adrenals: cytochrome b5 localization with respect to cytochrome P450 17,20-lyase activity. Mol Cell Endocrinol 2007;265–266:93–101.

36 Doi SA, Al-Zaid M, Towers PA, Scott CJ, Al-Shoumer KA: Steroidogenic alterations and adrenal androgen excess in PCOS. Steroids 2006;71:751–759.

37 Dunaif A: Insulin resistance and the polycystic ovary syndrome: mechanism and implications for pathogenesis. Endocr Rev 1997;18:774–800.

38 Brettenthaler N, De Geyter C, Huber PR, Keller U: Effect of the insulin sensitizer pioglitazone on insulin resistance, hyperandrogenism, and ovulatory dysfunction in women with polycystic ovary syndrome. J Clin Endocrinol Metab 2004;89:3835–3840.

39 Romualdi D, Guido M, Ciampelli M, Giuliani M, Leoni F, Perri C, Lanzone A: Selective effects of pioglitazone on insulin and androgen abnormalities in normo- and hyperinsulinaemic obese patients with polycystic ovary syndrome. Hum Reprod 2003;18:1210–1218.

40 Guido M, Romualdi D, Suriano R, Giuliani M, Costantini B, Apa R, Lanzone A: Effect of pioglitazone treatment on the adrenal androgen response to corticotrophin in obese patients with polycystic ovary syndrome. Hum Reprod 2004;19:534–539.

41 Romualdi D, Giuliani M, Draisci G, Costantini B, Cristello F, Lanzone A, Guido M: Pioglitazone reduces the adrenal androgen response to corticotropin-releasing factor without changes in ACTH release in hyperinsulinemic women with polycystic ovary syndrome. Fertil Steril 2007;88:131–138.

42 Azziz R, Ehrmann DA, Legro RS, Fereshetian AG, O'Keefe M, Ghazzi MN, PCOS/Troglitazone Study Group: Troglitazone decreases adrenal androgen levels in women with polycystic ovary syndrome. Fertil Steril 2003;79:932–937.

43 Pandey AV, Miller WL: Regulation of 17,20 lyase activity by cytochrome b5 and by serine phosphorylation of P450c17. J Biol Chem 2005;280:13265–13271.

44 Auchus RJ, Lee TC, Miller WL: Cytochrome b5 augments the 17,20-lyase activity of human P450c17 without direct electron transfer. J Biol Chem 1998;273:3158–3165.

45 Yanagibashi K, Hall PF: Role of electron transport in the regulation of the lyase activity of C21 side-chain cleavage P-450 from porcine adrenal and testicular microsomes. J Biol Chem 1986;261:8429–8433.

46 Lin D, Black SM, Nagahama Y, Miller WL: Steroid 17 alpha-hydroxylase and 17,20-lyase activities of P450c17: contributions of serine106 and P450 reductase. Endocrinology 1993;132:2498–2506.

47 Prough RA, Webb SJ, Wu HQ, Lapenson DP, Waxman DJ: Induction of microsomal and peroxisomal enzymes by dehydroepiandrosterone and its reduced metabolite in rats. Cancer Res 1994;54:2878–2886.

48 Waxman DJ: Role of metabolism in the activation of dehydroepiandrosterone as a peroxisome proliferator. J Endocrinol 1996;150(suppl):S129–S147.

49 Betz MJ, Shapiro I, Fassnacht M, Hahner S, Reincke M, Beuschlein F, German and Austrian Adrenal Network: Peroxisome proliferator-activated receptor-gamma agonists suppress adrenocortical tumor cell proliferation and induce differentiation. J Clin Endocrinol Metab 2005;90:3886–3896.

50 Kempná P, Hofer G, Mullis PE, Flück CE: Pioglitazone inhibits androgen production in NCI-H295R cells by regulating gene expression of CYP17 and HSD3B2. Mol Pharmacol 2007;71:787–798.

51 Lanzone A, Fulghesu AM, Guido M, Fortini A, Caruso A, Mancuso S: Differential androgen response to adrenocorticotropic hormone stimulation in polycystic ovarian syndrome: relationship with insulin secretion. Fertil Steril 1992;58:296–301.

52 Arslanian SA, Lewy V, Danadian K, Saad R: Metformin therapy in obese adolescents with polycystic ovary syndrome and impaired glucose tolerance: amelioration of exaggerated adrenal response to adrenocorticotropin with reduction of insulinemia/insulin resistance. J Clin Endocrinol Metab 2002;87:1555–1559.

53 Moghetti P, Castello R, Negri C, Tosi F, Spiazzi GG, Brun E, Balducci R, Toscano V, Muggeo M: Insulin infusion amplifies 17 alpha-hydroxycorticosteroid intermediates response to adrenocorticotropin in hyperandrogenic women: apparent relative impairment of 17,20-lyase activity. J Clin Endocrinol Metab 1996;81:881–886.

54 Storbeck KH, Swart AC, Slabbert JT, Swart P: The identification of two CYP17 alleles in the South African Angora goat. Drug Metab Rev 2007;39:467–480.

55 Munir I, Yen HW, Geller DH, Torbati D, Bierden RM, Weitsman SR, Agarwal SK, Magoffin DA: Insulin augmentation of 17alpha-hydroxylase activity is mediated by phosphatidyl inositol 3-kinase but not extracellular signal-regulated kinase-1/2 in human ovarian theca cells. Endocrinology 2004;145:175–183.

56 Rice S, Christoforidis N, Gadd C, Nikolaou D, Seyani L, Donaldson A, Margara R, Hardy K, Franks S: Impaired insulin-dependent glucose metabolism in granulosa-lutein cells from anovulatory women with polycystic ovaries. Hum Reprod 2005;20:373–381.

57 Stalvey JR: Inhibition of 3beta-hydroxysteroid dehydrogenase-isomerase in mouse adrenal cells: a direct effect of testosterone. Steroids 2002;67:721–731.

58 Pelletier G: Localization of androgen and estrogen receptors in rat and primate tissues. Histol Histopathol 2000;15:1261–1270.

59 Diamanti-Kandarakis E, Mitrakou A, Hennes MM, Platanissiotis D, Kaklas N, Spina J, Georgiadou E, Hoffmann RG, Kissebah AH, Raptis S: Insulin sensitivity and antiandrogenic therapy in women with polycystic ovary syndrome. Metabolism 1995;44: 525–531.

60 De Leo V, la Marca A, Lanzetta D, Cariello PL, D'Antona D, Morgante G: Effects of flutamide on pituitary and adrenal responsiveness to corticotrophin releasing factor (CRF). Clin Endocrinol (Oxf) 1998; 49:85–89.

61 Hines GA, Smith ER, Azziz R: Influence of insulin and testosterone on adrenocortical steroidogenesis in vitro: preliminary studies. Fertil Steril 2001;76: 730–735.

62 Abbott DH, Goodfriend TL, Dunaid A, Muller SJ, Dumesic DA, Tarantal AF: Increased body weight and enhanced insulin sensitivity in infant female rhesus monkeys exposed to androgen excess during early gestation (abstract P2-348). 89th Annual Meeting of the Endocrine Society, Toronto, 2007.

63 Liu J, Li XD, Vaheri A, Voutilainen R: DNA methylation affects cell proliferation, cortisol secretion and steroidogenic gene expression in human adrenocortical NCI-H295R cells. J Mol Endocrinol 2004;33: 651–662.

64 Weaver IC, Cervoni N, Champagne FA, D'Alessio AC, Sharma S, Seckl JR, Dymov S, Szyf M, Meaney MJ: Epigenetic programming by maternal behavior. Nat Neurosci 2004;7:847–854.

65 Goodarzi MO, Antoine HJ, Azziz R: Genes for enzymes regulating dehydroepiandrosterone sulfonation are associated with levels of dehydroepiandrosterone sulfate in polycystic ovary syndrome. J Clin Endocrinol Metab 2007;92:2659–2664.

66 Gambineri A, Pelusi C, Genghini S, Morselli-Labate AM, Cacclari M, Pagotto U, Pasquali R: Effect of flutamide and metformin administered alone or in combination in dieting obese women with polycystic ovary syndrome. Clin Endocrinol 2004;60:241.

67 McFadden D, Pasanen EG, Raper J, Lange HS, Wallen K: Sex differences in otoacoustic emissions measured in rhesus monkeys *(Macaca mulatta)*. Horm Behav 2006;50:274–284.

68 Herman RA, Wallen K: Cognitive performance in rhesus monkeys varies by sex and prenatal androgen exposure. Horm Behav 2007;51:496–507.

69 Ibáñez L, López-Bermejo A, del Rio L, Enríquez G, Valls C, de Zegher F: Combined low-dose pioglitazone, flutamide, and metformin for women with androgen excess. J Clin Endocrinol Metab 2007;92: 1710–1714.

70 Glueck CJ, Wang P: Metformin before and during pregnancy and lactation in polycystic ovary syndrome. Expert Opin Drug Saf 2007;6:191–198.

71 Salvesen KA, Vanky E, Carlsen SM: Metformin treatment in pregnant women with polycystic ovary syndrome – is reduced complication rate mediated by changes in the uteroplacental circulation? Ultrasound Obstet Gynecol 2007;29:433–437.

David H. Abbott, PhD
Department of Obstetrics and Gynecology and National Primate Research Center
University of Wisconsin, 1223 Capitol Court
Madison, WI 53715 (USA)
Tel. +1 608 263 3583, Fax +1 608 263 3524, E-Mail abbott@primate.wisc.edu

Author Index

Subject Index